T0258073

ON THE PROWL

ON THE PROWL

In Search of Big Cat Origins

MARK HALLETT AND JOHN M. HARRIS

COLUMBIA UNIVERSITY PRESS
New York

Columbia University Press
Publishers Since 1893
New York   Chichester, West Sussex
cup.columbia.edu
Copyright © 2020 Columbia University Press
All rights reserved

Library of Congress Cataloging-in-Publication Data
Names: Hallett, Mark, 1947- author. | Harris, John Michael, author.
Title: On the prowl : in search of big cat origins / Mark Hallett
    and John M. Harris.
Description: New York : Columbia University Press, [2020] |
    Includes bibliographical references and index.
Identifiers: LCCN 2019034905 (print) | LCCN 2019034906 (ebook) |
    ISBN 9780231184502 (hardcover) | ISBN 9780231545525 (ebook)
Subjects: LCSH: Panthera—Evolution. | Panthera—Origin.
Classification: LCC QL737.C23 H3326 2020 (print) |
    LCC QL737.C23 (ebook) | DDC 599.75/5—dc23
LC record available at https://lccn.loc.gov/2019034905
LC ebook record available at https://lccn.loc.gov/2019034906

Jacket art: On the mammoth steppe of the Trans-Caucasus Range of
western Russia, a pride of Eurasian steppe lions (*Panthera spelaea spelaea*)
begins an early morning hunt for reindeer or bison.

Cover design: Milenda Nan Ok Lee
Cover art: Mark Hallett

In memory of Dr. Alan Rabinowitz, whose sacrifices and hard work in understanding the biology of jaguars and other wild cats has provided hope for their continuance on Earth, and of Cathy McNassor, who never met a cat she couldn't bond with

The lions pass a thornbush and melt.
Though the whole day is unbroken
the passage of the sun will represent heaven
the bones will represent time.
   —Josephine Jacobsen

I feel
the link of nature draw me: flesh of flesh,
Bone of my bone thou art, and from thy state
Mine shall never be parted, bliss or woe.
   —John Milton

CONTENTS

ACKNOWLEDGMENTS

The authors wish to sincerely thank the people and institutions that helped make possible the writing, illustration, and production of this book. We especially wish to thank our scientific consultants: Dr. Xiaoming Wang and Dr. Blaire Van Valkenburg.

Dr. Wang is curator of vertebrate paleontology at the Los Angeles Museum of Natural History and an adjunct professor at the Chinese Academy of Sciences, the University of Southern California, and the University of California, Los Angeles. His areas of research focus on the evolution of fossil carnivorans, the late Cenozoic biostratigraphy of Inner Mongolia, and the mammalian evolution of the Tibetan Plateau and its paleoenvironments. He has published extensively on fossil canids and Asian mammalian paleontology; he and his associates discovered the oldest known pantherin, *Panthera blytheae*, during a 2010 excavation in southeast Tibet. Dr Wang reviewed the written and illustrated material relating to our treatment of the evolution of cold-adapted mammals of the Tibetan Plateau.

Dr. Blaire Van Valkenburgh is professor of ecology and evolutionary biology, curator of the Donald R. Dickey Collection of Birds and Mammals, and associate dean of academic programs in the Division of Life Sciences at the University of California, Los Angeles. She specializes in the study of large carnivoran mammals, and her recent work has focused on parallels between past and present carnivoran guilds, evolution of feeding adaptations, function and evolution of mammalian turbinates, and molecular and morphological evolution within the order Carnivora, Her publications include pivotal studies on the ice-age lions, sabertooths, and dire wolves of the Rancho la Brea fossil site. Dr. Van Valkenburgh reviewed our illustration concepts and text and provided us with current data relating to the overall predator-prey relations of large Asian and North American fossil carnivorans.

Drs. Marina V. Sotnikova and Irina V. Foronova, professors emeriti at the Russian Academy of Sciences, reviewed material relating to the anatomy and phylogeny of Pleistocene lions. The world-renowned wildlife biologist and conservationist Dr. George B. Schaller and his wife, Kay Schaller of the Wildlife Conservation Society, offered warm encouragement from the book's conception, reviewed and made invaluable suggestions to our treatment of pantherin conservation, and helped us find tiger consultants and data sources. Dr. Christine Argot of the Musée d' Histoire Naturelle, Paris, enabled us to photograph the Pleistocene lions in the museum collections. Kristine Loh and Danielle Garbouchian of the Panthera Corporation helped us obtain needed photographic material. Jun Huang's painstakingly detailed and accurate anatomical models of pantherins provided essential references in drawing. The photographers Bob Osburn and Ethan Crowley kindly permitted us the use of their photo images free of charge. Terri Pope scanned the original drawings. The production artist Karyn Servin turned the scans into master files for publication. Finally, we gratefully acknowledge Mark's fellow artist-paleontologist Mauricio Antón, whose knowledge, friendship, and superb renditions of ancient felids were an inspiration in illustrating and writing the book.

PREFACE

A young woman treads the narrow path rising from Silpuri village near the edge of Kanha National Park. Her feet pound softly, syncopating with the click of her ankle bracelets and producing clouds of dust that rise to mingle with the dawn air, barely cool from the stifling heat of the previous day. Shafts of light cut through the shadows of the nearby forest to illuminate tall grass along the path in crisp detail. Suddenly she halts, stops, her gaze locked on two golden, staring eyes inside a concentric maze of coppery-yellow, stark white, and black. The wide brown eyes of the woman are apprehensive; the golden ones of the female tiger, secure and mildly curious. Both are mothers. They are separated physically by a distance of several meters, but the sources of their reactions were crafted millions of years apart, and the passing seconds contain flashbacks of raw flesh, flint blade points, and bullets; of thrusting mountain strata and glacial cold; of gnawing animal and human needs; of royal grandeur, lifeless trophy heads, and now vanished forests. The tiger finally yawns, exposing a long, crenellated pink tongue and yellow, pointed canines, before calmly padding across the clearing and downhill to join her cubs at the streambed. Breathing out, the woman continues along the path to check on her family's rice paddy.

. . . . .

The great or big cats, known scientifically as the **pantherins** (terms in boldface type are explained in the glossary), are perhaps the most iconic of carnivores. Our instinctive fear of their lethal teeth, claws, and muscular power is balanced by an equally inborn fascination with their exquisite fur patterns, supple grace, and compelling presence. They have been represented for thousands of years on the walls of caves and palaces, on weapons, and in countless other images, ranging from loaded political symbols to innocent cartoons. As children we grow up with them as legends and myths, but only within the last sixty years or so, in the twilight of the pantherins' existence on our planet, have we started to begin to understand their behavior in the wild. We are now in the process of uncovering their origin, and this book will explore what is currently understood on the basis of new discoveries and current scientific interpretations. Although much is still unknown, enough clues have now been uncovered so as to create a tentative, sketchy trail of where the great cats came from, sometimes crossing paths with our own emergence as modern humans. The story of the great cats' evolution is based firmly in the fossil record, new interpretations of comparative anatomy, and the decades of patient field studies by wildlife biologists. Where the trail becomes faint, however, we must occasionally resort to intelligent, informed speculation, admitting what we do not know. This new summation in turn provides the basis for how we can work to ensure that pantherins and other wild cats will continue into the future.

The late British paleontologist Alan Turner (Liverpool John Moores University) and the Spanish artist-paleontologist Mauricio Antón (Museo Nacional de Ciencias Naturales) in 1997 jointly produced a superbly presented, highly readable, and beautifully illustrated book, *The Big Cats and Their Fossil Relatives* (Columbia University Press), which recounts what was known at that time about cat evolution, focusing in particular on both the true felid and nimravid "sabertoothed" cats. This was followed by the excellent *Sabertooth* (Indiana University Press, 2013), by Antón, and *Biology and Conservation of Wild Felids*, by David MacDonald and Andrew Loveridge (Oxford University Press, 2010). These two titles are classic works of paleontology and zoology, and because of the thoroughness with which the nimravid and machairodontine sabertoothed cats have now been assessed, their subject matter is therefore generally outside the scope of our book. Instead, we focus on the evolution and paleobiology of tigers, jaguars, leopards, lions, and other pantherin cats within the broad context of ancient changing ecosystems and on their possible relationship with other animals and humans, including the first documented contacts during more recent prehistory.

Chronologically, our story of big cat evolution focuses on the last 23 million years of the Earth's history, an interval of time that contains the Neogene and Quaternary **periods**. The Neogene comprises two **epochs**. The Miocene epoch started at 23.03 million years ago or, as abbreviated by earth scientists, at 23.03 **Ma**. The subsequent Pliocene epoch started at 5.33 Ma. The Quaternary period contains the Pleistocene epoch, which began at 2.59 Ma, and the Holocene epoch that began about 10,000 years ago (~10 **ka**) and has lasted until the present day. The epochs are in turn subdivided into **stages** or **ages** (appendix 1). The stages of the epochs are, for historical reasons, defined by different sequences of marine sedimentary rocks located in Europe; an age is the interval of time that it took to accumulate that rock sequence or stage.

Most fossil mammals occur in terrestrial rocks, and it is not always easy to precisely match or correlate European sequences of terrestrial rocks with their marine chronological equivalents; correlating the terrestrial rocks of other continents with the European marine stages is even more difficult. Mammal paleontologists have, therefore, subdivided the terrestrial fossil-bearing sequences of the different continents into their own unique sequences of **land mammal stages (LMS)** and **land mammal ages (LMA)**. Each stage has been assigned a type locality and is defined by its unique assemblage of mammals. For example, the type locality of the North American **Rancholabrean** stage is the Rancho La Brea Tar Pits of Los Angeles, and the Rancholabrean LMA begins with the arrival of bison into North America, recently established by Froese (University of Alberta, Edmonton) and colleagues as about 195,000 years ago. The largest fossil cat known from complete skeletal material, the American lion *Panthera atrox*, occurs only in the Rancholabrean LMA. In North America, the Neogene LMAs most important to our story are the **Hemphillian** (10.3–4.9 Ma) and the Rancholabrean (240 ka–11ka; see appendix 2).

Much of our story follows big cat history in Europe and Asia. The first mention of pantherin animals (species, families, etc.) that are of particular importance will also appear in **bold** and are listed in appendix 4. When describing the different events, we will—where possible and for clarity—use the epoch names and dates in millions of years. Where we need to be even more precise, we'll provide the European or Asian stage names. For Europe, the most important of these for our story are the **Villanyian** (3.6–2.58 Ma), **Gelasian** (2.58–1.81 Ma), and **Calabrian** (1.81–0.78 Ma). The Asian LMAs most important to our story are the **Mazegouan** (3.6–2.59 Ma) and the **Nihewanian** (2.59 Ma–700 ka). Our emphasis on Asia, a crucial region for the origin of big cats, reflects the extensive but generally less-well-known discoveries and academic studies over the last fifteen to twenty years in Asian paleontology by Chinese and other workers.

In addition, the timeless Middle and Late Paleolithic paintings and portable sculptures of pantherins in caves such as Chauvet-Pont-d'Arc and others in Europe offer a unique insight into the impact these cats had on the human mind in these places and times, and these are examined in chapter 5, "Testimony of the Caves." We also seek to reconstruct the past ecologies of the pantherins' endangered and, in some cases, now lost subspecies, such as the Caspian tiger and Barbary lion, beginning with the first records of the animals and ending with a survey of present-day surviving populations. The historic relationship between ethnic cultures and great cats, combined with the most recent field studies of pantherins, offers the possibility that new conservation attitudes and practices could bring these animals back from their currently endangered status to once again become wild, free carnivores. It is the authors' fond hope that this book will in some way contribute to this outcome.

ON THE PROWL

CHAPTER 1
Threads in the Fabric of Time

Gujarat, India, 2020: Hungry and beset by flies on a hot afternoon under thorny acacia trees, a small pride of Asiatic lions endures its discomfort. Once ranging over vast territories spreading across the Mediterranean lands and western Asia, this pride and the other scattered groups in and around the Gir Forest are now all that are left (figure 1.1).

. . . . .

The Asiatic lions (*Panthera leo persica*) are only the tiniest end threads of a vast and patchy but continuous evolutionary fabric stretching back millions of years. The apocalyptic fireball that ended the dinosaurs' enormously long domination of land ecosystems at the end of the Cretaceous period 65.5 million years ago left a world of unfilled niches to be exploited by small mammalian survivors and others. Most of the earth was still locked in a "hothouse" climate, with temperatures well above those of today, and, although the drifting continents had mostly assumed positions close to those they occupy today, the gigantic raft forming the subcontinent of India had not yet collided with Asia, which at that time was bisected by a wide seaway—the Turgai Strait—running north to south. During the Paleocene epoch Europe, like modern-day Indonesia, was a group of islands of various sizes that were close to and at some places geographically connected by two main land bridges at the northeastern area of Canada (figure 1.2). The humid, shadowy tropical forests of the newly separated Europe and North America, though dominated by smaller arboreal mammals, were soon to be exploited by larger terrestrial,

---

FIGURE 1.1. Asiatic lion pride stratigraphic sequence with fossils.

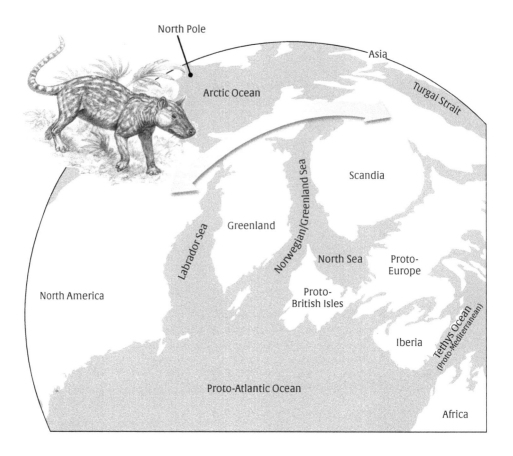

FIGURE 1.2. North Atlantic land areas during the Late Paleocene–Early Eocene. During the early Tertiary (60–55 Ma) the land masses that would become North America, Greenland, and Europe were much closer than they are today, facilitating the dispersal (arrows) of early carnivorans (such as *Dissacus* sp., *left*) and other mammals.

forest-floor species. Some of these were still relatively unspecialized and had teeth that could handle a variety of foods, both plant and animal, and they included the jackal-sized, mesonychid *Dissacus europaeus* and the wolf-sized arctocyonid *Arctocyon primaevus*. Although they were omnivores (*omnis*, "all" + *vorare*, to devour), they showed signs of beginning to specialize in carnivory (flesh eating); other species specialized in frugivory (fruit eating), folivory (leaf eating), or insectivory (insect eating). Flesh is metabolically easier and quicker to digest than plant food, so at this early period, as potential prey types diversified, there was an incentive for some forms to evolve toward this lifestyle. Here we should make a distinction between a **carnivore** (*any* animal that eats flesh) and a **carnivoran**—a

FIGURE 1.3. Skull of miacid, *Miacis cognitus* (also see plate 1).
The skull of *Miacis cognitus* had cheek teeth with pointed cusps that were effective in shearing flesh. Its long cranium served as the anchor for a powerful temporalis muscle, giving this marten-sized carnivoran a powerful bite.

mammal belonging to the order Carnivora, the one that includes all living, terrestrial flesh-eating mammals but also some omnivorous (dogs), herbivorous (giant pandas), and insectivorous (aardwolf) forms. Thus, *Tyrannosaurus rex* was a carnivore but not a carnivoran, and the giant panda is a carnivoran but not a carnivore.

The first carnivorans appeared some 50 million years ago in the late Paleocene epoch. They were mostly weasel sized (~30 cm or 11 in.), with long, slender bodies, short limbs, long tails, and low, shallow heads, and belonged to the **Miacidae** family (of the superfamily Miacoidea). "*Miacis*" means "tiny pointed" and describes the occlusal surfaces of their teeth (figure 1.3). Each of their sharp, five-toed claws could retract upward along the outside of the toe, suggesting an early adaptation for both climbing and seizing small arboreal and forest floor prey (plate 1). The miacids' prey, like that of modern pine martens and civets, may have included the earliest rodents, marsupials, small birds, and reptiles. Later (middle Eocene) forms like European *Paroodectes feisti* (figure 1.4)—whose beautifully preserved skeletons have been found in localities like Messel, Germany—show that the miacids, although successful, remained only small, arboreal predators. Why not larger, more diverse ones?

The miacids were early members of the (as yet unranked) clade Carnivoramorpha and became part of the carnivore community by the late Paleocene. However, for many million years they were overshadowed by the **creodonts** (*kreas*, "flesh" + *odons*," tooth"), a diverse, highly successful, and long-lived order that also began in the Late Paleocene and flourished until the early Miocene. Characterized by wolf-sized members

FIGURE 1.4. Skeleton and life restoration of *Paroodectes feisti*.
About the size of a housecat and possibly like a raccoon or civet in its habits,
the middle Eocene miacid *Paroodectes* was adapted for climbing and jumping in
the tropical forests of the Proto-European archipelago, where it pursued insects,
reptiles, and small mammals and also ate fruit. This species is known from a
complete skeleton (*upper left*) preserved in the Messel Pit fossil locality southeast
of Frankfurt, Germany, and shows great similarity to miacids from the Bridger
Formation of North America.

like *Hyaenodon horridus* (figure 1.5), creodonts included a wide range of
fully carnivorous species that filled many of the niches occupied by later
modern carnivorans such as weasels, wolves, hyenas, and bears. Creodonts
may have arisen from Paleocene or possibly even late Cretaceous insecti-
vore specialists, the **paleoryctids**. In this group, the central parts of their
upper and lower tooth rows—the ones that originally might have been
good at crushing insects—evolved into shearing blades that could cut flesh.
Whether or not this was the origin of the creodonts' shearing innovation,

FIGURE 1.5. Restoration of *Hyaenodon horridus*.
Advanced creodonts like the wolf-sized *Hyaenodon* were the main terrestrial carnivores for millions of years and prevented the earliest carnivorans, the miacids (seen in the tree above), from radiating beyond the role of small, arboreal predators. The skull of *Hyaenodon* skull is shown in figure 1.6.

such teeth provided a definite evolutionary advantage for efficiently processing flesh as food. Accordingly, creodonts got a head start over the miacids and other early carnivorans and remained the dominant small-to-large-bodied terrestrial mammal carnivores until the latest Oligocene. *Paroodectes* and other small miacids persisted through the Eocene, like minor actors waiting in the wings offstage for their chance at a larger role, while the smaller creodont species diversified into genet- to raccoon-sized hunters.

The same evolutionary adaptation can arise more than once, however, in totally unrelated animals. Early unspecialized **eutherian mammals** ("true beasts") had four premolars and three molars in their upper and lower tooth rows. In the creodonts, the first upper molar and the second lower molar (but in some forms, the second upper molar and the third lower molar) performed the flesh-shearing function, and for this reason are called **carnassial** teeth. The early carnivorans independently evolved their own carnassials, but in their case the shearing blades were formed from the last upper premolar and first lower molars. Thus, the carnivorans retained molars behind the first carnassials that were still capable of processing other foods by crushing or grinding. That the creodonts' rear-most molars were modified as carnassials meant that they lacked more posterior molars for processing other foods. The posterior molars of miacids and some of their much later descendants (like canid carnivorans) were available for processing nonflesh items, which would prove advantageous when prey species were scarce (figure 1.6).

Their limb structure and mobility might also have given carnivorans a competitive edge. The feet of many (but not all) creodonts were held in a **plantigrade** position, i.e., the soles of the feet were placed flat on the ground, whereas those of the carnivorans were mostly **digitigrade**, bearing the weight on the toes. The creodont carpal (wrist) and tarsal (ankle) structure was less mobile than that of carnivorans. As a result the creodonts were less agile hunters than the carnivorans, whose more efficient joints allowed them to move more swiftly.

Ultimately, the creodonts' fate may have been sealed by climate and temperature changes. A dramatic shift in climate starting about 55 Ma is indicated by a change in the shape of fossil leaves. As in today's forests, the ancient leaves from warm, rainy climates have smooth margins and pointed "drip-tips" to sluice away water and prevent molding, whereas those of cooler, drier climates are smaller and have serrated margins because water accumulation is not a problem. At the end of the Eocene, about 34 Ma, the proportions of these leaf types changed. In North America, Europe, and Asia warm, year-round temperatures of 22°C (72°F) had produced humid evergreen (laurellaceous) forests. These were replaced by temperate, deciduous, broad-leafed (sclerophyllous) forests as the mean annual temperature lowered by 12°C (22°F). As the temperatures became progressively cooler and drier, the extensive forests gave way to drier and more open habitats, resulting in a corresponding change in the herbivores that depended on these plants for food. More open habitats meant that herbivores had to rely

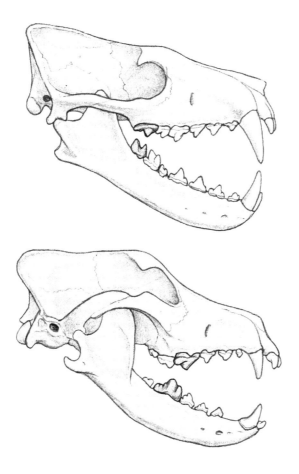

FIGURE 1.6. Comparison of the skulls of *Hyaenodon* and *Canis*.
The carnassials or shearing teeth of *Hyaenodon* (*above*) and the modern wolf *Canis* (*below*) are shown in gray. Although modern mesocarnivores like wolves prefer flesh, their most posterior molars also enable them to consume other foods that *Hyaenodon*'s specialized carnassial teeth would have been unable to effectively process.

on speed rather than shelter to escape their predators. The lesser mobility of the creodonts was a competitive disadvantage even though, at this point in time, the teeth of the carnivorans were less specialized.

Because mammals as a group have highly specific feeding adaptations, dental morphology is especially useful in determining their evolutionary relationships. We know, on the basis of their dental morphology, that by the Late Eocene carnivorans had separated into two major superfamilies, the **Arctoidea** (or Canoidea) and the **Aeluroidea** (or Feloidea). The arctoids are superficially bearlike or doglike, and their extinct and living members include the ursids (bears), canids (dogs), and mustelids (weasels and relatives), among others. The arctoids' cheek teeth display a generalized chewing capability for handling meat but in some types retain a capacity for crushing or grinding plants, and for this reason many arctoids are termed **mesocarnivores**, or eaters of flesh in moderate amounts. The more catlike aeluroids include groups like the civets, mongooses, and hyenas as well as the cats (figure 1.7). Aeluroids differ from arctoids by displaying greater specializations for carnivory. Around 40 Ma during the mid-Oligocene the aeluroids had diversified into six families, and one of these groups was the **Felidae family**—the felids or cats. First becoming distinct from the two closely related families, **Nimravidae** and **Barbourofelidae** (see later in this chapter) they continued to separate, along with the civets and relatives (**Viverridae**), first, from the mongooses and meerkats (**Herpestidae**), then from the viverrids, and finally from the hyenas (**Hyaenidae**). Felids became specialized as **hypercarnivores**, or extreme meat-eating specialists. In cats, the enlarged first premolar of the upper jaw and first molar of the lower jaw, the carnassials, became latterly compressed, sharp ridges that could efficiently slice through the toughest muscle and tendon. As part of this extreme specialization other teeth, like some of the other premolars and molars, became less and less functional, eventually becoming tiny vestiges or disappearing altogether (see chapter 2).

Along with specialized cheek teeth for processing flesh, the felids also evolved another distinctive feature, a shorter face that resulted in a more powerful bite. A short face brings the upper and lower canine teeth, the main instruments for making a kill, closer to the compressive force that closes the jaws. In all mammals the force closing the jaw is supplied mainly by two powerful, paired jaw-closing muscles, the masseter and temporalis. These work together by contracting from opposite directions to pull the jaw tight. The masseter originates on the zygomatic arch (cheek bone) and attaches to the outer side of the mandible or lower jaw, while the temporalis originates on the broad expanse of the side of the skull and attaches to the top of the mandible's coronoid process. The advantage of a predatory mammal with a short skull versus a long-jawed one can be compared to the greater force generated by a pair of short grasping tongs versus that of longer ones—the pressure created by the arms of the former is stronger (figure 1.8). This bite power is exemplified by that of the tiny weasel (*Mustela nivalis* and others), which, although weaker than a felid's in absolute strength, is comparatively much stronger for its size because of the short face and longer surface on the skull to which the temporalis attaches. For felids, a shorter muzzle left less room for the large internal olfactory surfaces that give arctoid hunters such as wolves and bears such a keen

FIGURE 1.7. Phylogeny of the carnivorans.
During the early Oligocene epoch the order Carnivora separated into two suborders, the Aeluroidea (or Feliformia), which included the cat, hyena, and civet families, and the Arctoidea (or Caniformia), which included the dog, bear, and weasel families and their relatives. The precise relationship of these families is still a matter of debate among phylogenetic workers. (animals not to scale)

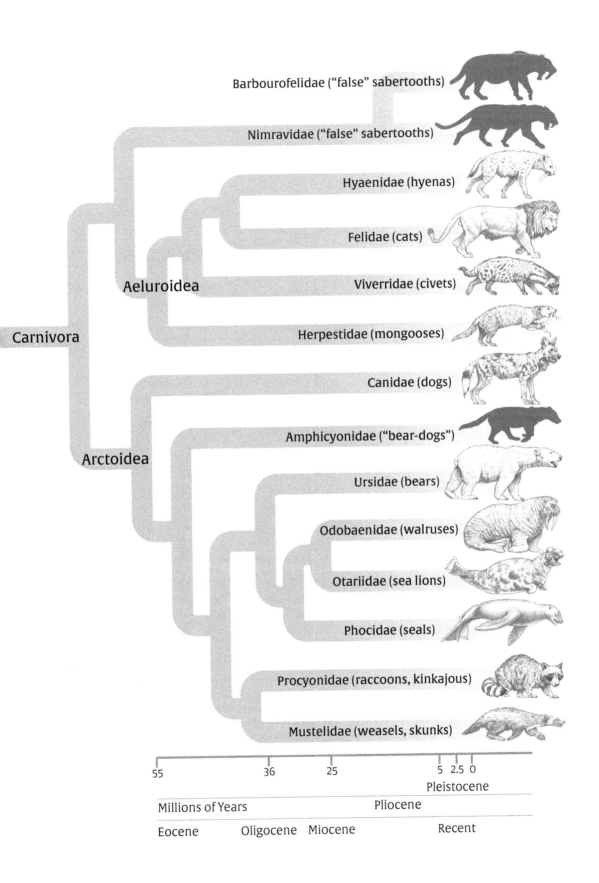

Barbourofelidae ("false" sabertooths)

Nimravidae ("false" sabertooths)

Hyaenidae (hyenas)

Felidae (cats)

Viverridae (civets)

Aeluroidea

Herpestidae (mongooses)

Canidae (dogs)

Carnivora

Amphicyonidae ("bear-dogs")

Arctoidea

Ursidae (bears)

Odobaenidae (walruses)

Otariidae (sea lions)

Phocidae (seals)

Procyonidae (raccoons, kinkajous)

Mustelidae (weasels, skunks)

55   36   25   5 2.5 0

Pleistocene

Millions of Years   Pliocene

Eocene   Oligocene Miocene   Recent

Size of fox and weasel skull compared (scale 1:5)

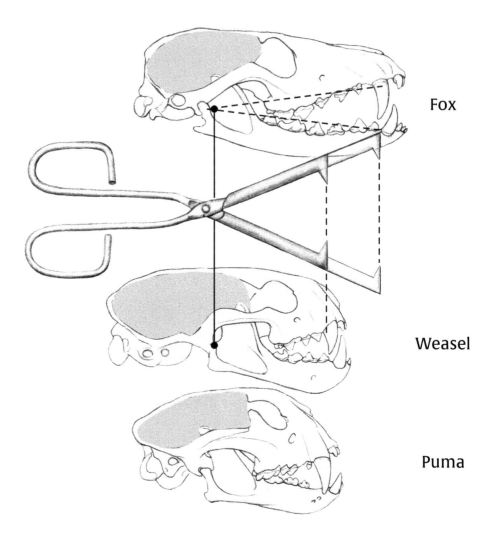

Fox

Weasel

Puma

FIGURE 1.8. Jaw strength comparison.
Although much smaller in size and absolute jaw strength, a weasel like *Mustela nivalis* can produce, for its size, a comparatively far greater amount of bite force than that of the larger red fox, *Vulpes vulpes* (shown to scale, *above*). As with the ends of a pair of short tongs, the meeting of teeth closer to the pivot point creates more force than those set farther away. Shorter jaws in which the canines meet closer to the pivot point (black dots) produce more power than the longer ones. A weasel's longer cranium also provides more anchorage for a strong jaw-closing muscle, the temporalis, which also gives it a proportionately more powerful bite that can kill a rabbit far bigger than it. These two factors are what give a felid like a puma (*below*) more jaw power than a wolf of similar size when killing a deer.

sense of smell, but this was compensated for by increased visual acuity. The elimination of a long muzzle also enhanced the felids' binocular vision by increasing the amount of overlap in the lower visual field; excellent nocturnal vision is characteristic of all cats and gives them an advantage over other carnivorans in judging prey distances during night-time hunts.

In addition to a formidable bite, the earliest felids also possessed highly mobile forelimbs. The ability of the forepaws to rotate medially and outward (enabled by the lateral movement of the radius, one of the two lower arm bones), combined with extendable claws, allowed the restraint of moving prey before a killing bite was administered (figure 1.9). In this respect felids generally became committed to an ambush mode of predation rather than to cursorial (running) predation as practiced by later canids and cheetahs. In dogs, speed in pursuit was achieved by reducing the mass (and therefore weight) of lower limb musculature and by evolving long legs to increase stride length in running. In felids, however, the more robust limb musculature required for strength in grappling and overcoming solitary prey meant greater weight. Although powerful musculature made possible sudden, rapid acceleration for ambushes, it also compromised sustained speed, except in the living African cheetah *Acinonyx jubatus* and cheetah-like fossil forms like the North American *Miracinonyx trumani* (or *Puma* [*Miracinonyx*] *trumani*), whose limbs are more doglike in proportion. Cats also generally lack another important quality, stamina, which is achieved by supplying the skeletal muscles with sufficient oxygen to maintain high, prolonged activity levels. This requires a large volume of highly oxygenated blood flow from the heart. A wolf's or African Cape hunting dog's amazing ability to eventually exhaust and overcome a running deer or antelope after a long, protracted chase is attributable in part to their having a proportionately much larger heart and lungs. By contrast, although a cat of similar size to these dogs can pursue its prey over a distance of about 182 meters (200 yards) at speeds of about 48 kph (30 mph), it then tires quickly and if unsuccessful usually gives up the chase.

The earliest felids were probably similar to the primitive living fossa (*Cryptoprocta ferox*) or Malagasy civet (*Fossa fossana*) from the island of Madagascar—agile, treetop hunters of small vertebrates. With the decline of the creodonts the available niches for predatory hypercarnivores were finally open for the felids to exploit. In spite of their adaptive advantages, however, felids were confronted early in their evolution by a *fifth* aeluroid family, the Nimravidae, which paralleled the morphology and probable behavior of the felids in many remarkable ways. The nimravids, or "false sabertooths" were originally (but incorrectly) given the name **paleofelids** ("ancient cats") to distinguish them from the **neofelids** ("new cats" or true felids) since at one time they were considered to be the ancestors of the neofelids. Nimravids, however, differ from true felids in a minor

FIGURE 1.9. Tiger using forepaws to capture prey.
Big and small cats use their forepaws, which they can turn inward and outward, to surround and take down small, agile animals; here a tiger captures an Asian mouse deer, *Tragulus kanchil.*

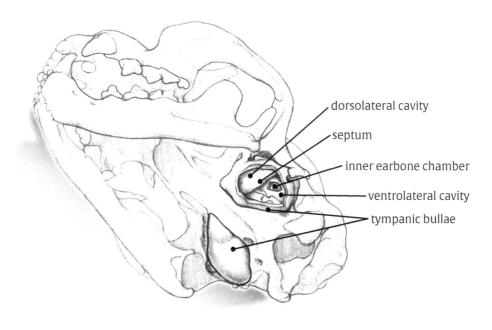

dorsolateral cavity

septum

inner earbone chamber

ventrolateral cavity

tympanic bullae

FIGURE 1.10. Auditory bullae.
A cat skull in ventrolateral view shows the paired auditory bullae (heavy outlines), hollow bony chambers that surround the three auditory ossicles or inner ear bones. The ossicles actually transmit sound waves to the auditory nerve and brain. The hollow, resonant chambers of the bullae (*cut view at right*) amplify fainter sound waves, in carnivorans aiding in the detection of prey. Unlike in nimravids, the bullae of felids are divided completely by a bony septum.

but fundamental way that makes an ancestor-descendant relationship impossible. All carnivoran families possess **auditory bullae** (also called tympanic bullae; see figure 1.10), paired bony inflations at the underside and rear of the skull that partly encapsulate and protect the middle ear bones. In true felids the interior of each bulla is divided into an inner and outer chamber by a bony septum, but in nimravids this separation and sometimes the entire bulla was absent or was only cartilaginous. Lasting from late in the Oligocene (36 Ma) until about the beginning of the Pliocene (5 Ma), nimravids are considered to be evolutionary convergents or look-alikes with respect to felids. Convergents are unrelated species that develop an almost identical morphology and behavior. Nimravids ranged in body size from small, probably arboreal hunters, such as the ocelot-sized *Dinictis squalidens* to jaguar-sized species like *Pogonodon platycopis* (figure 1.11). A closely related but separate family, the Barbourofelidae, included huge, massively powerful members like *Barbourofelis fricki*, capable of taking down rhinos. Many or even most

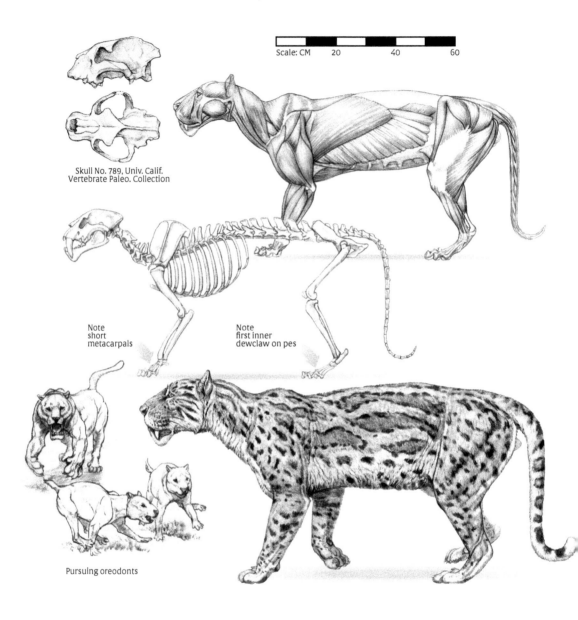

Scale: CM   20   40   60

Skull No. 789, Univ. Calif.
Vertebrate Paleo. Collection

Note
short
metacarpals

Note
first inner
dewclaw on pes

Pursuing oreodonts

FIGURE 1.11. Restoration of *Pogonodon platycopis*.
Although the skeleton of a fossil animal may be fragmentary, a thorough study and
comparison of related forms can aid a paleoartist in making a *reconstruction* of the
skeleton, as Mark Hallett has done here with *Pogonodon*, a Late Oligocene "false
sabertooth" nimravid from the John Day Formation of eastern Oregon. About
the size of a modern jaguar, *Pogonodon*'s known limb proportions may have been
similar to that of its smaller, earlier relative *Dinictis squalidens*, whose skeleton is
much more complete. The patterning and fur texture here are conjectural, but
the artist has used those of a similarly sized large modern pantherin to show the
animal's original appearance in life, a process known as a *restoration*.

nimravids were amazingly similar to the **machairodontines**, the felid subfamily of "true" sabertoothed cats, with which they competed. The nimravids' persistence in so many forms and over such a long geological period made them serious competitors with the felids throughout most of their existence.

Details of the overall early evolution of the felids or "true cats" are still very patchy and not well known. The likelihood of a dead animal becoming a fossil is very low, but carnivorans' individual numbers are usually far fewer than those of their prey. As a result, among a given fossil fauna, carnivorans may well have been present in an ancient environment but are less well represented than more abundant herbivores. Exceptions do occur, such as the famous Late Pleistocene "La Brea Tar Pits" (Rancho La Brea locality) of the Los Angeles Basin, which preserved literally thousands of the smilodontin *Smilodon fatalis* fossils, and the late Pliocene Bolt's Farm locality near Johannesburg, South Africa, where several specimens of the metailurin *Dinofelis barlowi* were found in a sinkhole. Both of these are "natural trap" situations where predator remains actually outnumber those of prey because of an unusual situation: trapped prey lured predators that in turn became unable to escape and died in place. These are examples of **preservation bias**, in which a particular type of animal or plant may be under- or overrepresented in the fossil record because of unusual conditions of accumulation.

In order to be preserved as a fossil, an animal normally has to be buried under anoxic (oxygen free) conditions, such as in river, lake, or ocean sediments. An arid, mountainous region may teem with life, but because the habitat is primarily an erosional rather than a depositional setting, fewer fossils will be preserved than, for example, in an alluvial plain. The latter might have only a modest species diversity but could accumulate a wealth of skeletal remains, some from animals that never actually lived there, because of the frequent rate and amount of sedimentation caused by flooding. Forests have acidic organic soils that often dissolve bone, while the alkalinity of more arid habitats is more likely to preserve it. All these factors affect the preservation of fossils and hence what we know about ancient cats.

For a moment, let's go down the rabbit hole of taxonomy (*taxis*, "to move" + *nomen*, "to name") and **systematics** to see where felids fit within our human classification system. The placement of the cats still follows the system of assigning names proposed by the Swedish botanist Karl von Linné, or Linnaeus, in the mid-eighteenth century. Linnaeus proposed giving all life forms a **binomial**, or two-named designation based on their distinctive anatomical traits or characters—also known as binomial nomenclature. The first word in the name (the **genus**, plural **genera**—always capitalized) is the equivalent of a human last name or surname, and the second (the **specific name**—never capitalized) is like a human first name. Together they single out the organism from all others in a precise way. Although other closely related creatures can have the same genus name, none can have the same species name. The living lion, for example, is **Panthera leo** (*Panthera*, "pantherin cat"; *leo*, "lion"), while the common leopard is **Panthera pardus** and the jaguar **Panthera onca**. Sometimes a third name is added after the species epithet to denote a distinct subspecies or race: **Panthera tigris** is the tiger, but **P. tigris altaica** refers to the rare Amur ("Siberian") tiger, while **P. tigris sumatrae** is the even rarer Sumatran tiger. Sometimes a genus is divided into two or more **subgenera**; a **subgenus** name appears in parentheses after the genus name but before the species epithet. As a result some specialists assign a subgenus status to the extant savanna lion, **Panthera (Leo) leo** (see chapter 4), while the genus name for the American fossil cheetah, *Miracinonyx*, is regarded by some workers as a subgenus of *Puma*, hence *Puma* (*Miracinonyx*) *trumani*. Although the similar-sounding Latin names may at times seem confusing, they serve to pin down the nature of the animals we want to understand: snow leopards and clouded leopards aren't really leopards at all, and opinion about the true identity of the giant ice age pantherin cat of North America, *Panthera atrox*, has swung back and forth between a jaguar and a lion.

While taxonomy attempts to clarify and pin down the *identity* of an organism, **phylogeny** (*phyllos*, "group" + *genesis*, "origin") seeks to unravel *relationships*; this science is at the heart of the search for where felines, large and small, living and extinct, came from. If we think of all mammals as being like the growth on a tree or bush, then the felines and other living mammalian carnivoran families are all branches of a big limb, the order Carnivora (a group known specifically as "carnivorans," as opposed to the general term "carnivore" for any flesh-eating animal). At the branch's fringe our lion species, *Panthera leo*, would be just a bud, along with buds for the other big cat species, on the twig of the genus *Panthera*. All of the pantherin species are what are known as a crown group—closely related species or genera that radiated from a common ancestor within a relatively short geological period of time. A common ancestor and all its descendants form a **monophyletic** ("single lineage") relationship known as a **clade** (*clados*, "branch"), and scientists refer to these systems as **phylogenetic** trees, based on a comparison of the physical differences between one species and another.

What exactly, then, is a pantherin? Simply put, of course, they are the "big" or "great cats," for most of us, supersized versions of housecats. The word "**panther**" may originally trace back to ancient Sanskrit for tiger, *pudarikam*, but probably derives from the Greek *pan* ("encompassing" or "all") and *therion* ("beast," implying a mammal), and the word may have originally referred to the leopard, which was still common in territories dominated by the ancient Greeks. The Latin version, *panthera*, was adopted by the German naturalist Lorenz Oken in 1816 to describe one genus within Felidae, the cat family. A century later, in 1917, the British taxonomist R. I. Pocock proposed, based on the basis of shared cranial features, that only the lion, tiger, leopard, jaguar, and snow leopard (but not the cheetah, the puma, or other popularly known big cats) should be formally included in the genus *Panthera*. In spite of this, these animals were until the mid-twentieth century sometimes assigned to the genus *Felis* or (in the case of lions) *Leo*.

The classification of the big cats, although sometimes still a point of disagreement among specialists, has now more or less reached a consensus. Pocock's classification of *Panthera* as a genus still holds, and this genus name is the derivation of the **Pantherini** tribe, which includes all the species of this genus (figure 1.12). Members of this tribe are in this

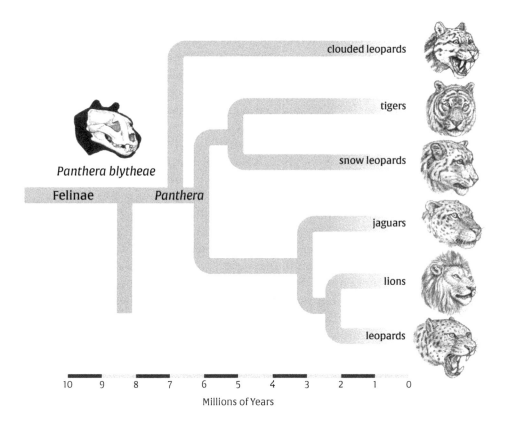

FIGURE 1.12. Phylogeny of the pantherins.
Following the origination of the clouded leopard group (*Neofelis*) about 7 Ma, the tribe Pantherini appears to have arisen at least 5.95 Ma with the appearance of *Panthera blytheae*, a basal form lying at or close to the clade's origins. The snow leopards and tigers of Asia are closer to each other anatomically and genetically than either is to the jaguar-lion-leopard group, which may have its origins in either Africa or Asia about 3 Ma. (animals not to scale).

book therefore are referred to as "**pantherins**." (Some workers recognize a "pantherine lineage," based on molecular studies, which includes not only the genus *Panthera* but also a number of small to midsize living felines, for which they propose a separate subfamily, the **Pantherinae**; if this becomes accepted then the term "**pantherines**" for its member genera would be correct.) The second so-called big or great cat forms, the puma and its close relative, the jaguarundi, and the cheetah, are placed in a separate tribe, the **Acinonychini**. A number of small-cat genera like lynx, ocelot, clouded leopard, margay, and others constitute the **Felini** tribe, and these three branches in turn form the subfamily **Felinae** (note the "n"). The Felinae are balanced by the Machairodontinae subfamily, the extinct "true" saber-toothed cats (although some workers, such as Per Christiansen [Aalborg University], do not recognize this as a monophyletic clade), and together the two subfamilies make up the Felidae (note the "d")—the entire cat family, living and extinct. In short, all members of the living big and little cats belong to the Felinae and are therefore **felines**. With the exception of the snow leopard, all pantherins differ from other cats in having an elastic ligament within the slender hyoid bones of their throats underneath the skull. This ligament stretches when the cat vocalizes, producing the characteristic "roar" of lions and tigers, and may have evolved as an aid in vocalizing to proclaim territory (figure 1.13).

Comparative anatomy, a discipline that interprets the morphological similarities or differences of animals both living and extinct, is the basis of our understanding of feline and later big cat origins. Their interpretation is based largely on structures observable to the eye, like those of bones or soft tissue. With fossil cats, skeletal remains can be damaged or incomplete—if you have only one bone or tooth specimen that just slightly differs from another it might be a new species, but it could also just reflect the natural variation that occurs within a population of a single species. To make an informed decision, it helps to compare the fossil specimen with as large a sample of closely related species as possible because large samples help us see patterns of similarity. Pertinent characters of the sample are measured and entered into a database. The more specimens you have in your sample (and hence the larger the data set), the greater is the likelihood that you'll be able to interpret these relationships correctly.

Extinct cats are usually represented by bones and teeth, and the differences in shape from those of other cats, backed up by precise measurements, are often the basis of determining whether a particular fossil form might have been an ancestor, descendant, or close relative of another. A common tool in comparative anatomy is **morphometrics**, the precise study and measurement of skeletal features. When linear measurements taken from skulls and other bones that have diagnostic value are subjected to multivariate analysis, the resulting data allow comparisons to be made

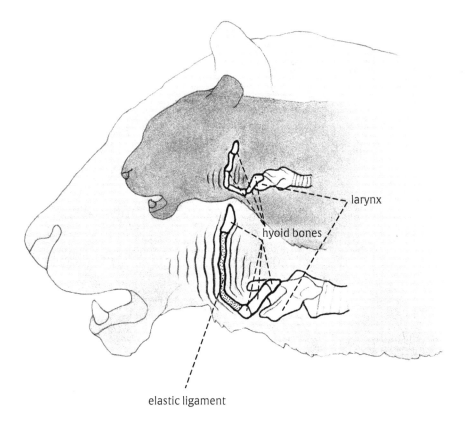

FIGURE 1.13. Hyoid bones and sound.
In small cats a series of small, paired, interconnected bones, the hyoid bones, are located at the base of the tongue, where they serve as a source of attachment for the muscles that aid in swallowing. These in turn attach to the larynx or voice box, and when a cat purrs, the air passing through the larynx causes the hyoids to vibrate, which amplifies the sound. In pantherins, however, the bones are connected at one point by a ligament, whose elastic properties can create much more intense vibrations, producing these cats' characteristic roar.

between living and fossil forms, which can shed light on evolutionary relationships. Dental characters, especially those of the premolars and molars, are especially important since the shape, presence, or absence of the cusps and other structures can help to tell us, when geological ages are factored in, who's an ancestor and who's a descendant.

In evaluating relationships paleontologists often consider two conditions, **basal** and **derived**, terms that have replaced the earlier "primitive" and "specialized." This is because a supposedly primitive species may actually have advanced, highly evolved features yet, compared to its descendants, represent an earlier, more generalized stage of overall development.

In cats we encounter this in the clouded leopard *Neofelis nebulosa*, a living feline whose cranial features place it near the base of the pantherin evolutionary lineage but whose hyperlong canines are specialized for penetrating the deep fur or feathers of its arboreal prey. A derived species that evolved from a basal one may have highly developed features like this, but in organic evolution there's no "end goal," as in the old, erroneously linear depictions of fossil horse or early-primate-to-human lineages. A later form *may* be more specialized than its ancestor but is not necessarily so; a derived species is one that's simply later in time than a basal one.

Within the last few decades two relatively new disciplines have come to the aid of traditional comparative anatomy for interpreting cat evolution and relationships. One is **cladistics**, which, like comparative anatomy, studies similarities and differences but instead of attempting to initially build phylogenetic trees and track possible evolutionary routes at a macro level, deals with the number of **shared characters** or traits at a micro level. In cladistics the focus is on what and how many characters two or more species may have in common. There may be many shared characters, and various computer programs and mathematical disciplines are used to evaluate or "weigh" their relative importance in relation to one another. Several concepts can help determine relationships. From these a **cladogram**—a diagram of two or more species' possible relationships with each other and their derivation from a common ancestor—can be constructed. It's still based on anatomical comparisons, but here the approach is more quantitative rather than the traditional qualitative or somewhat subjective interpretive method. This helps to determine whether two or more forms naturally group together based on shared characters, which in turn provides a basis for constructing phylogenetic trees. As logically precise as cladistics may seem, however, this method can and often does produce disagreement among paleontologists, who may arrive at differing interpretations of the same characters and then construct differing phylogenetic relationships. The concepts of plesiomorphy, synapomorphy, and autopomorphy are sometimes used to sort out which characters may be used to establish the relationship of one species with another. Plesiomorphy (*plesio*, "near" + *morphos*, "shape"), means that an ancestral characteristic is shared by more than one clade or group of animals; an example of this would be the partially or totally divided tympanic bullae that all felids share with their aeluroid ancestors. Two or more animals, sharing one or more characters that only these two have in common to the exclusion of all others, are said to have a synapomorphy (*syn*, "shared"); if only *one* has a particular feature it's called an autopomorphy (*auto*, "self").

A second, more recent science of understanding biological relations is **molecular phylogenetics**, in which hereditary cellular materials like DNA, RNA, and other proteins are studied to show evidence of changes in the **genomes**, or genetic codes, of populations and individual species over time. In theory this produces a biochemical record of evolutionary lineages independent from those phylogenetic trees inferred only from the anatomy of fossils. Although molecular phylogeny arrives at conclusions about relationships by different means than cladistics, it essentially takes the same approach in that its findings seek to verify that such relationships are based on shared characters and that all species have a common origin. Here genetic materials, often **mitochondrial DNA (mtDNA)**, are extracted and isolated from the frozen cells of blood or skin tissue of living animals (and sometimes fossil ones), if necessary subjected to chemical amplification (cloning or replication) for easier study, and analyzed by genetic sequencing for comparison. For example, in a 1995 study of pantherin group origins, Dianne Janczewski (Frederick Cancer Research and Development Center) and colleagues amplified, sequenced, and compared two chromosome base-pair regions, the 12S ribosomal RNA (12S) and the cytochrome b coding gene (Cytb) from seventy-five individual animals, representing seventeen felids and five nonfelid carnivorans. From the sequencing results, they were able to assemble clusters of data that enabled some important relationships among the different cat groups to be determined. To ensure accuracy, multiple scans were made from several individuals within each species, along with subspecies and, in some cases, populations within a species. One reason for such extensive cross-checking is that some characteristics of genes can produce misleading results. Among these are **long branch attractions** (LBA), where long evolutionary lineages can falsely appear to be closely linked because of accumulated genetic change within one of them; saturation, a false reading in the apparent divergence rate between two lineages, again because of accumulated genetic changes, common in fast-evolving species; and transitions, reversions to previous characters in some base-pair genes. Establishing a large database by the this kind of cross-checking minimized problems in the Janczewski study, and as a result her team came to three basic conclusions: First, the genus *Panthera* evolved relatively recently, is monophyletic, and is represented by six extant species—the lion, leopard, jaguar, snow leopard, tiger, and clouded leopard. Second, the African cheetah and the puma share a common ancestry. And third, two golden cat species—the Asian golden cat (*Felis temmincki*) and African golden cat (*Profelis aurata*)— are not sister species (two closely related species recently derived from a common ancestor), and the African species is more closely related to the caracal (*Caracal caracal*). Nevertheless, despite this convincing body of evidence—and as with cladistics—a worker who places more importance on a species' first fossil occurrence may arrive at conclusions different from those of a molecular phylogeneticist as to the evolutionary length of the lineages and when the splits in lineages, or speciation events, occurred. Ideally the phylogenetic trees of both disciplines should agree, but they can sometimes differ (figure 1.14).

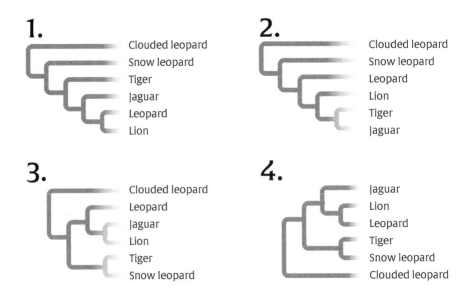

FIGURE 1.14. Differing cladogram models of pantherin phylogeny. Cladistic analysis is based on a study of derived characters, those that appear in recent parts of a lineage but not in its older forms and that can be then be used to construct a cladogram, a diagram that resolves the evolutionary relationships among organisms like pantherin cats. Unlike other types of phylogenetic trees, however, cladograms make no implications about the time or the relative amount of character change, and because their methods of analysis (biometric vs. molecular or the "weight" or importance assigned to certain characters, for example) can differ, researchers may produce differing cladograms, as shown here. Cladogram 4 is presented as the most correct depiction of extant pantherin phylogeny by Davis, Li, and Murphy (2010).

Understanding the duration of evolutionary lineages and when speciation events occurred, in effect creating a "biological calendar," is a major goal of molecular phylogenetics. Sequencing from both Cytb *and* mtDNA is done because Cytb appears to accumulate genetic differences relatively rapidly and then stabilize, whereas those of mtDNA's 12S accumulate gradually but steadily. This makes Cytb a good indicator of sister species that have recently split from each other, possibly no more than a few million years ago, whereas the 12S is more apt to cast light on moderate to longer phylogenetic divergences. The characteristics of these molecules suggest that the fairly large number of closely related felid species found in some regions of the world (ten in South America alone) was probably in some cases caused by rapid species-radiation events within a short geologic time period (ocelots, margays, Chilean cats, pampas cat, and others in the genus *Leopardus*, for example), but others (pumas, genus *Puma*, and jaguar, genus

Panthera) took much longer. It also helps to understand another important consideration: biogeographical distribution—the geographic distribution of a species or group of species. Felids are highly mobile and seem to
have rapidly colonized habitats in many areas, although their origins are
not always clear. Members of the *Leopardus* clade, for example, could have
either arisen within a certain area of South America to later fan out across
the continent or, alternatively, might have dispersed to South America from
North America after the Panamanian land bridge arose about 2–3 Ma.

In evaluating relationships of fossil felids over time much importance
was until recently given to the inferred overall body size of a given individual specimen. A concept known as the Cope-Depéret Rule states that
among many clades of animals there is a tendency for overall size increase
through time, which relates to overall adaptability. Although this is generally true of felids, size does not in itself imply a direct phylogenetic relationship, and it's important to heed the warnings of J. H. Mazák (Shanghai
Science and Technology Museum) and Andrew C. Kitchener (National
Museums Scotland, Edinburgh), both tiger specialists. They and their colleagues point out that individual members of a cat species can vary greatly
in size (sometimes because of gender or ecological factors) and that overall size is not necessarily a good diagnostic factor. This is especially true
when only one specimen of a new species is known.

The first solid anatomical evidence for a felid-resembling species
dates back at least 25 million years with **Proailurus lemanensis** (figure 1.15,
plate 1b) and possibly as early as 28.5 Ma with very basal "cats," *Stenogale
intermedia* and *Haplogale media*. *Proailurus* is known from two individuals at
the early Miocene (22 Ma) location of Saint-Gérand-le-Puy, France, and
also from other, currently undescribed cat fossils dating from 17–16.5 Ma
in the Nebraskan Miocene deposits at the Ginn Quarry, East Cuyumungue,
Sheep Creek, and Echo Quarry in Nebraska. These fossils are sometimes
informally referred to as a "*Proailurus* grade" (rather than as the genus *Proailurus*) to denote that they represent a stage in a felid evolutionary trend.
Proailurus, as one would expect at this early point, was a small animal, with
a skull about 15 cm (5.9 in.) long and, as reconstructed by Antón, was
about 38 cm (15 in.) tall at the shoulder. The vertebrae are not well preserved, but the limb bones are. These and fossil material from other sites
suggest an animal similar to and slightly larger than the living fossa and,
at about 9 kg (20 lb.), adept at hunting both arboreal and ground prey.
At this stage, the *Proailurus* skull shape and tooth morphology are mostly
modern in form, but the teeth are more numerous, not yet having undergone the reduction characteristic of more modern cats. Although some
specialists place *Proailurus* as only a highly derived viverrine on the way
to becoming a felid, others, such as Lars Werdelin (Swedish Museum of
Natural History), on the basis of the basicranial area of the skull and teeth,
consider it to be the first true felid.

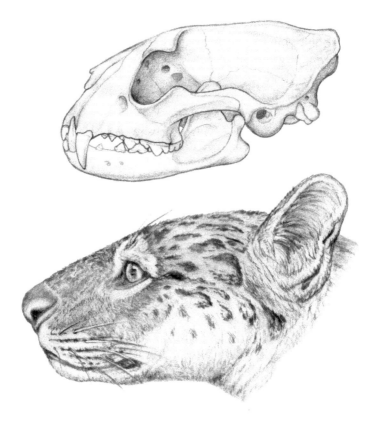

FIGURE 1.15. Skull and restoration of *Proailurus lemanensis.*
Somewhat larger than a domestic house cat, *Proailurus lemanensis* had excellent
arboreal abilities and a lifestyle probably similar to the living fossa of Madagascar.
A complete skull is known from the Ginn Quarry, Nebraska, which displays the
longer muzzle and unreduced dentition (also see plate 1).

Eocene-Oligocene localities in the Quercy region of France have
yielded most of the material of early European carnivorans older than
21 Ma, but felids other than *Proailurus* are surprisingly poorly known.
A similar situation occurs in North America, where some workers informally refer to an interval of time 25–18.5 Ma, (early to mid-Miocene) as
the "Cat Gap" because of the scarcity or actual absence of fossils.

However, felids representing this time interval have now been found
from the Ginn Quarry and other Nebraskan localities, and it is possible
that the "Cat Gap" may reflect a change in habitat distribution. Beginning
at about 25.8 Ma a cooling trend—initiated by the opening of the
circum-Antarctic seaway and accompanied by changes in North Atlantic
ocean currents and glaciations in Antarctica, Greenland, Iceland and
southern Alaska—resulted in the gradual retreat of the formerly widespread

humid, closed-forest ecosystems in much of the world, including North America. These were replaced in part by fragmented, open forested areas and in many areas by widespread **savannas**, in which cooler-adapted and arid-adapted grasses and shrubs predominated, with fewer trees. An evolutionary explosion of new mammalian herbivores followed, with many arctoid carnivores such as canids, ursids, mustelids, and amphicyonids coevolving to prey on them. These open environments tended to favor more cursorial (running) carnivores such as dogs over ambush predators such as cats, which preferred the dense vegetation habitats that favored ambush hunting. This cause-and-effect hypothesis, proposed to explain the "Cat Gap," is debated among paleontologists, but it should be remembered that although the Miocene's general cooling-and-drying trend may have created more open-country habitats suited to fast-running herbivores and the cursorial carnivorans that preyed upon them, closed-forest habitats never entirely disappeared. Forest conditions are not favorable for bone preservation, and this could explain the lack of ambush-hunting felid fossils.

In the early Miocene we encounter another felid, **Pseudaelurus tournauensis** (= *Pseudaelurus transitorius*, figure 1.16), which is thought to have directly evolved from the Late Oligocene–Early Miocene *Proailurus* or "*Proailurus* grade" forms. First known from Wintershof-West, Germany in deposits dating from 20–18 Ma, *Pseudaelurus* lived in Europe, the Arabian peninsula, and Asia from about 20 Ma and reached North America by about 18.5 Ma, the first undisputed felid genus to do so. Here it's known by *P. validus* from the late Hemingfordian–early Barstovian (17–15 Ma) and from the smaller *P. stouti* and *P. skinneri* from the Hemphillian of Hawk Rim, Oregon (10.3–4.9 Ma). Although considered by some specialists as a true monophyletic group, others refer to *Pseudaelurus*, as in the case of the "*Proailurus* grade" Nebraskan fossil felids, as an informal stage or "grade" in felid evolution. Others place *Pseudaelurus* into other genera: *Styriofelis*, *Hyperailurictis*, *Miopanthera*, or *Schizailurus*. *Pseudaelurus* was housecat sized, and its relatively short legs suggest that it, too, was an arboreal hunter. Although *Pseudaelurus tournauensis* was small, it evolved to exploit larger-cat hunting niches, progressing from the lynx-sized, European *P. lorteti* to *P. quadridentatus* that, at around 30 kg (66 lb.), was almost as large as a cougar. The basal *P. tournauensis* persisted in Europe until 8 Ma, but the two larger forms died out 2 million years earlier. *P. romieviensis*, smaller than *P. tournauensis*, is also known from Europe, and *P. guangheensis* and *P. cuspidatus* have been recovered from China. The *Pseudaelurus* "grade," however, is important for another reason. Paleontologists sometimes refer to the felines as the "conical toothed" cats to distinguish them from their extinct relatives, the "sabertoothed" machairodontine cats, whose upper canines were usually compressed from side to side. The *Pseudaelurus* grade, appears to have been ancestral both to the machairodontines or felid sabertooths as well as the "conical-toothed" felines, the pantherins and

FIGURE 1.16. Skull and restoration of *Pseudaelurus turnauensis.*
Pseudaelurus was a diverse genus that was widespread during the early Miocene and probably evolved directly from the Oligocene *Proailurus.* Several forms, ranging in size from that of a housecat (*P. turnauensis*) to that of a puma (*P. quadridentatus*), are known from Africa, Eurasia, and North America, and some authorities favor splitting the genus into several separate genera. The importance of *Pseudaelurus* lies in its position as a common ancestor of both the conical-toothed cats (subfamily Felinae) and the sabertoothed cats (subfamily Machairodontinae).

other modern living cats. North American *Pseudaelurus,* while possessing the "bow tie" shape of the mandibular m1 tooth's paraconid and protoconid that is characteristic of true felids (see chapter 2), also shows more advanced features in comparison with Eurasian species, such a shorter total tooth row length. Significantly, while the canines of other *Pseudaelurus* are circular in cross-section, possibly presaging the conical-toothed felines, the canine of *P. skinneri* is oval in cross-section, anticipating the mediolaterally compressed canines of the machairodontine cat line.

The dominance of the last nimravids and barbourofelids and the early machairodontines as specialist predators of large prey probably kept the conical-toothed felids confined to the role of smaller predators for many millions of years. In the Turolian stage of the late Miocene, between 8.7 and 4.9 Ma, however, there was an evolutionary explosion of small- to medium-sized cats that diverged from the ancestral *Pseudaelurus* lineage. Paradoxically, this actually occurred *before* a geologically later, widespread radiation of new rodent types that, had it occurred earlier or at the same time as the cats', would have neatly explained the latter's success because of the widespread new prey base. But instead we have a paleontological mystery.

The first divergence from the ancestral *Pseudaelurus* lineage led to a group that comprised the clouded leopards and the pantherins at about 10.8 Ma (figure 1.17). The clouded leopard *Neofelis*, with its hyperlong canines and distinctive skull features, soon split off from the pantherins about 7.0 Ma, and, by about 4 Ma, gave rise to today's two species, *N. nebulosa* and *N. diardi*. Another divergence close to the *Panthera-Neofelis* split is that of the *Pardofelis* or bay cat clade, typified by species like the Asian golden cat (*Pa. temmincki*), which separated about 9.4 Ma. This was followed by the *Caracal* lineage at 8.5 Ma, which incorporated medium-sized African cats such as the caracal (*C. caracal*) and the serval (*Leptailurus serval*); the ocelot lineage at 8.0 Ma, made up of the ocelot (*Leopardus ocelot*) and mostly other small exclusively South American cats; and also the genus *Lynx*, which separated at 7.2 Ma to give rise to both the New and Old World species (*Lynx canadensis, L. rufus* and *L. pyrenaica*). It also gave rise to the medium to large cats of the tribe Acinonychini, which appeared at 6.7 Ma and eventually evolved into the small rain forest puma or jagua-rundi (*Puma yagouaroundi*), the mountain lion or cougar (*Puma concolor*), the extinct North American cheetahs (*Miracinonyx trumani* and *M. inexpec-tatus*, classified by some specialists, who regard *Miracinonyx* as a subgenus, as *Puma [Miracinonyx] trumani* and *Pu. [Miracinonyx] inexpectatus*), and the extant Old World cheetah (*Acinonyx jubatus*; see later in this chapter). Finally there arose the small cats of the Old World, which became an entity at about 6.2 Ma. All belong to the genera *Felis* and *Pristifelis* except for the Pallas's (flat-headed) cat (*Otocolobus manul*) and the leopard cat (*Prionai-lurus bengalensis*).

The earliest recognized species of the genus *Felis* used to be *Felis attica*, but that species was recently transferred to the genus *Pristifelis* following a reevaluation of its dental characters. The genus *Felis* is now restricted to more recent species such as *Felis margarita* (the sand cat), *F. sylvestris* (European wildcat), *F. lybica* (African wildcat), and *F. chaus* (Asian jungle cat). *Pristifelis attica* (figure 1.18) is a late Miocene cat found in several European fossil localities from Spain to Greece. It differs from *Pseudaelurus* by lacking the second lower premolar and the posterior cusp of the second

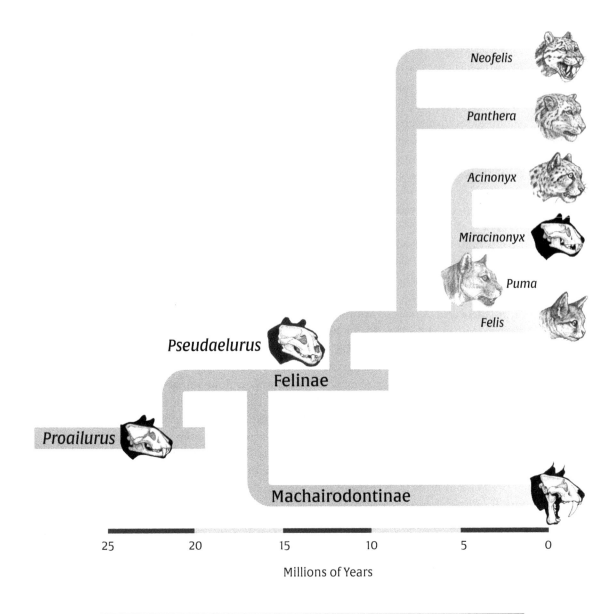

Millions of Years

FIGURE 1.17. Phylogeny of the Felinae and Machairodontinae.
The basal felid *Pseudaelurus* is currently considered a potential ancestor to both
the sabertoothed cats (subfamily Machairodontinae), first known during the very
late Oligocene to Early Miocene, and the conical-toothed cats (subfamily Felinae),
which appeared later in the Miocene. The Felinae include the tribe Felini that
comprises the living "small cats" of the Old and New World, the pantherin lineage
(tribe Pantherini), and the pumas and cheetahs (tribe Acinonychini). (Animals
not to scale.)

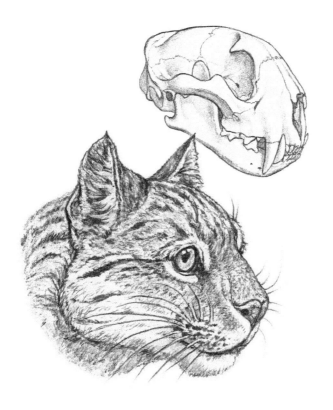

FIGURE 1.18. Skull and restoration of *Pristifelis attica.*
Known from the Late Miocene of about 12 million years ago, this bobcat-sized
species is an early representative of the most diverse group of today's extant cats.
It still retains a small upper first premolar (P1), which is absent in modern members
of this genus.

lower molar (see chapter 2) and was evidently ancestral to the extant representatives of the genus *Felis.*

Along with the small hunters, other forms evolved that began to compete as medium-sized predators with the last of the nimravids and later felid
sabertooths. These are the previously mentioned tribe Acinonychini, or
pumas and cheetahs. They, as well as the homotherins ("scimitar-toothed
cats," sometimes shortened to "homotheres"; see chapter 4), had a role in
competing with the early Pleistocene pantherins and a discussion of their
origins is appropriate here. Members of the tribe Acinonychini, puma-like
cats, diverged from other felines during the late Miocene at about 6.7 Ma,
with the highly specialized cheetahs later splitting off from the more basal
jaguarundi-puma lineage in the early Pliocene at 4.9 Ma. Like jaguars,

acinonychins *may* have had an Asian origin; fragmentary but diagnostic middle Pliocene fossils are known from the Transcaucasian area of central Asia and Mongolia. But as Darren Naish (University of Southampton) and others point out, the group could alternatively have arisen from puma-like ancestors that dispersed to North America. As noted in chapter 4, Helmut Hemmer (Johannes Gutenberg Universitat, Mainz) considers the African Laetoli fossils formerly attributed to pantherins as belonging not only to pumas but perhaps even to the modern genus *Puma*. This could mean that acinonychins (as with jaguars, leopards and lions; see chapter 4) could have had an *African* origin, later dispersing to Europe, then Asia, and finally to the Americas.

An extinct South American puma species, *Puma pumoides* (originally described under the name *Felis pumoides*, and not to be confused with *Pu. pardoides*), is known from cranial and postcranial fragments found in the probably Pliocene "Brocherense Bed" stratum in the Reartes Valley of the Córdoba Province, Argentina. These fragments definitely have jaguarundi-like (and therefore acinonychin) characters and establish the arrival of pumas in the New World early in pantherin history. *Puma pumoides* may have given rise not only to the extant cougar *Pu. concolor* but also to the Pleistocene North American cheetah *Miracinonyx*.

Formerly described and named as variously *Felis* (later *Panthera*) *pardoides* (Owen's panther) and *Veretailurus schaubi* (Schaub's panther), these previously diagnosed Early Pleistocene felines from Europe with alleged leopard- or lion-like affinities were shown by Hemmer in 2004 to be conspecific and so puma-like that they are now referred to *Puma pardoides*, the "Eurasian puma" (figure. 1.19). Similar to *Pu. concolor* in its shortened face, long back, and relatively short forelimbs but lighter at 40–45 kg (88–99 lb.), *Pu. pardoides*, like the modern species, was probably highly adaptable in its choice of habitats but may have preferred lightly wooded areas, making it yet another competitor with early jaguars, leopards, and *Megantereon* sabercats.

Basal cheetahs had become characteristic fast-pursuit predators of the Asian steppes by the late Pliocene–early Pleistocene and were present on the African savannas and, as forms convergent with "true" Old World cheetahs, on the American steppes. The Asian *Sivapanthera linxiaenensis* (figure 1.20), was probably a direct ancestor of the later Nihewanian (early Pleistocene) *S. pleistocaenica* and competed with "cursorial" hyenas like *Chasmaporthetes progressus* for swift antelopin prey that may have included the ancestors of the modern Tibetan chiru (*Pantholops hodgesoni*) and the saiga (*Saiga tatarica*). In North America, pumas gave rise to the genus *Miracinonyx* (as mentioned already, considered by some as a subgenus of *Puma*), that was highly convergent with the Old World cheetahs. At least two valid species are known, *Miracinonyx trumani* and *M. inexpectatus*, and their skeletal features indicate that, like the Eurasian and African forms, these were adapted for high-speed pursuit, certainly the equal of North America's

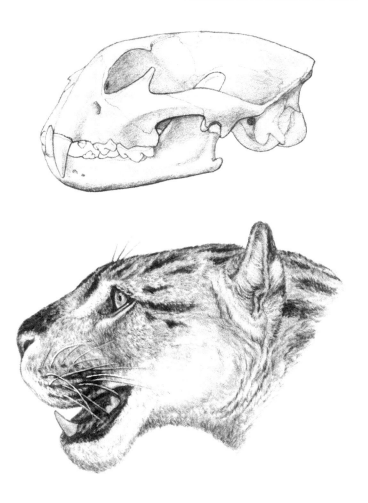

FIGURE 1.19. Skull and life restoration of *Puma pardoides*.
Originally referred to a variety of different, early pantherin species from the
Villanyian LMS of the early Pleistocene of Italy and from several different
localities, the short facial region and other characters of this skull has led some
workers to recognize these cats as representatives of European pumas.

sole surviving pronghorn antelope, *Antilocapra americana*. The American
cheetahs, once considered to have shared a common ancestor with the Old
World forms, are now considered, on the basis of genetic studies, to have
evolved in parallel but separately. Because of its greater skeletal similarity to
cougars, *M. inexpectatus* is considered to be a more basal form of the genus
than is *M. trumani*. In Eurasia, species of the extant cheetah *Acinonyx* had
evolved by the Villanyian LMS (late Pliocene). A giant form, *Acinonyx pardinensis*, is known both from a complete skull recovered from the French
site of Saint Vallier and from a vertebral column and limb bones at the

FIGURE 1.20. Early cheetah pursuing prey.
The early cheetah *Sivapanthera linxiaenensis* chases a sheep, *Ovis yushenesis,* across a Chinese grassland. During the Gaozhuangian Asian LMS ancestral cheetahs like *Sivapanthera* (skull, *lower right*) evolved to prey on the cold-adapted plains ungulates and were probably the direct ancestors of the modern cheetah.

slightly older locality of Perrier in the Massif Central. Lion-sized and, at about 80–100 kg (175–220 lbs), twice the weight of the modern *A. jubatus,* its skeletal proportions, although long and slender, are more like those of a snow leopard than of a modern cheetah (figure 1.21). Perhaps it was not quite as fast as *A. jubatus* is in pursuing small and medium-sized antelopes, but it nevertheless would have been an effective predator in running down big, fast-moving bovids that were too speedy for ambushing lions or homotheres.

As threatened as they are today by habitat loss, hunting, and poaching, the highly diverse, cosmopolitan small cats have survived until the present, an indication of their successful adaptation to a variety of environments ranging from hot, dry deserts to humid, wet rainforests. Hypercarnivorous hunters of small game, they evolved to occupy the predatory niches that

FIGURE 1.21. Skull and life restoration of the giant cheetah, *Acinonyx pardinensis.* During the early and middle Pleistocene a cheetah that reached the size (but not weight) of a modern lion lived on the steppes of Eurasia. Although a swift cursorial hunter, its less-specialized inner ear structure, body proportions, and somewhat heavier build suggest that it was slower and less adept at fast turns than a modern cheetah. The giant form is compared with the extant species, *Acinonyx jubatus,* at lower right.

the nimravids and machairodontines ignored in favor of larger prey. The *Neofelis-Panthera* clade was ancestral to the pantherines or "big cats" we know so well, but how and when did the pantherin cats become the species now familiar to us? How could they compete in the role of large-bodied feline hunters when the sabertooths, the master predators of large game, already monopolized this predatory niche? We will discover the answer to this shortly, but to understand the evolutionary emergence of the pantherines we first need to examine the anatomy and behavior of their small ancestors to find out what makes them such remarkable predators.

.   .   .   .   .

In the Gir thorn forest the sun has now burned out into a red, sullen dusk and, in the growing darkness, the lions are stirring in readiness for another hunt . . .

CHAPTER 2
Anatomy of a Hunter

Saint-Gérand-lePuy, France, 22 million years ago: Stretched out on a shaded branch in lonely, spotted splendor, an eight-month-old *Proailurus lemanensis* cub surveys the windy, game-dotted plain. Driven away by the increased unwillingness of his mother to share her kills and the appearance of an aggressive adult male in his native woodland, his survival in this new territory will depend on his ability to put his acquired hunting skills to use (figure 2.1).

. . . . .

By the early Miocene the family Felidae had emerged as among the most specialized of land predators. We have already discussed some main features that made them so efficient at carnivory: prey-grappling claws that could be kept sharp by retracting into the toes; a shortened muzzle that produced a powerful, concentrated bite; and highly compressed carnassial teeth that could efficiently shear through and process flesh. Anatomy and behavior evolve in tandem, however, so let's look at how the bodies and behaviors of a felid combine to make them such effective hunters.

The most distinctive aspects of felid dentition are the **premolars** and **molars** (or cheek teeth) that reflect how a prey animal is processed after being killed. The lower first molar has a "bow tie" shape that is characteristic of true felids (figure 2.2). Cats never evolved the ability to crack bones for the nutritious marrow, as a canid or hyaenid does, although some larger species occasionally chew on the ends of bones such as ribs and scapulae and some ingest quantities of bones when swallowing small prey. The shearing carnassials (last upper premolar and first lower molar) became especially high ridged and laterally compressed, like a knife blade,

---

FIGURE 2.1. *Proailurus* surveying a floodplain in France during the Late Oligocene.

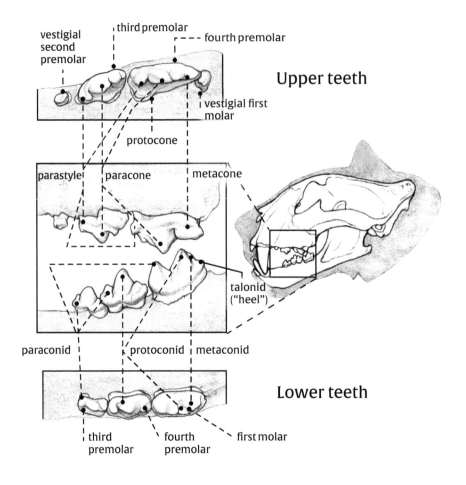

FIGURE 2.2. Felid dentition.
A tiger's left upper and lower molars, or chewing teeth, in (a) upper occlusal view, (b) lateral view, and (c) lower occlusal view. As in other modern cats, some teeth characteristic of the basal carnivoran set of premolars and molars are absent or vestigial. When precisely named and measured, the teeth provide a basis of comparison among individuals of different species, allowing researchers to ascertain relationships between living and extinct forms.

suitable only for slicing. This explains why a housecat (*Felis sylvestris domesticus*) will sit and appraise its food before shearing off a corner through the side of its mouth (figure 2.3). It's a careful and instinctive way of dealing with meat, as opposed to a domestic dog (*Canis lupus familiaris*), which simply grabs a piece of meat with its incisors and canines and immediately moves the food to the back of the mouth for chewing. Cats, like most carnivorans, may also use their **incisors** for biting off flesh close to the bone, when the larger teeth can't be used easily, and the papillae (keratinized, backward-pointing protuberances) of the tongue's

FIGURE 2.3. How a cat consumes flesh.
Cats process flesh in three basic ways: (a) shearing food with the rear carnassial teeth, (b) pulling meat away from a carcass with the canines and incisors, and (c) rasping meat from a bone with the concave, keratinous papillae of the tongue's upper surface (*inset*).

upper surface can also efficiently remove meat from a bone simply by licking it over and over. Because cats developed such specialized carnassial teeth for shearing soft flesh, eventually several of the other upper and lower premolars and molars became less functional; these eventually became reduced in size and in some cases were totally lost. This evolutionary pattern of changes in the morphology of the cheek teeth plays an important role in evaluating fossil pantherin relationships and evolution, as we will see in chapter 4. As hypercarnivores, felids also process soft flesh quickly in their gut systems—so quickly that sometimes some of it fails to be thoroughly digested. This is why Egyptian vultures (*Neophron percnopterus*) have been observed eating lion feces for their nutritive value and why domestic dogs, the ultimate gluttons, can't seem to be broken of their habit of raiding cat litter boxes to eat their contents.

A felid's cusped incisor teeth, as already mentioned, not only play a role in helping the canines to pull meat away from a carcass but also, like the canines, assist to a limited extent in holding or gripping to subdue live prey. In advanced machairodontines like *Homotherium* there's no doubt of this function: the incisors are massive, almost canine-like, and project well forward. Both the upper and lower incisor rows in these cats are spaced so that they effectively interlock, and the pointed ends ensure enough penetration to hold. The incisors of modern felines are far less impressive but nevertheless perform a vital function in assisting the canines to remove chunks of flesh and in acting like combs for grooming the fur.

The canines are the instruments that do the actual killing. A canine's tapered shape, with apex directed downward to channel the maximum amount of force and concentrate it at a single point, has throughout time remained the basic design for cats and other carnivores to penetrate flesh. Nimravid and machairodontine upper canines, however, are laterally compressed with points or *denticles* along the *carina* (keel) or cutting edge. These serrations, like those of steak knives, assist in slicing flesh. Sabertooths evolved to kill by means of a specialized **shear-bite**, so let's compare this method of dispatching prey with that of a conical-toothed feline.

Sabertoothed cats—the nimravids, barbourofelids, and machairodontines—independently evolved their hyperenlarged upper canines to exploit large, sometimes tough-skinned herbivores. Although most carnivorans will take both large and small prey, larger animals that can be overpowered and killed provide more food and hence are a better "payoff" for the amount of effort and calories expended. At the same time there are dangers inherent in attacking large herbivores, which can defend themselves by rending with tusks, biting, fatally goring, or giving a disabling kick, sometimes breaking the attacker's jaw and ensuring its death by starvation. Every second that a killing is prolonged increases this kind of risk, and the speedier the kill the better. Smaller animals can be dispatched with a bite to the back of their skull, by breaking the neck or piercing the spinal cord, but larger, more robust prey like a buffalo presents a bigger challenge (see chapter 4). By the middle Miocene the larger-bodied felid predatory niche was already occupied by the last representatives of the nimravids and their close relatives the barbourofelids, both of which, along with the early machairodontines, had developed the specialized shear-bite (canine shear bite). Although some workers (Fabrini 1890) had previously alluded to this process, William A. Akersten (Idaho State University, Pocatello) in 1985 offered the first detailed hypothesis of this mechanism, which countered the popular (but erroneous) "stabbing" concept of earlier authors. Highly developed in later forms, the shear-bite was initiated after the cat had used its powerful forelimbs to wrestle the prey off its feet. With the prey held down securely, the cat firmly held the end of its lower jaw against the neck while opening its jaws wide so that the upper and lower canines aligned on opposite sides of the prey animal's throat. The immensely strong neck muscles pulled the sabertooth's skull down so that the long, laterally compressed upper canines ("sabers") could perform the kill. While the lower jaw acted as a stabilizing brace to aid their directional force, the upper canines arced down and back to pierce the throat, well below the prey's neck vertebrae and tough upper muscles. The serrated, posterior concave edges of the upper canines moved downward, slicing past the serrated, convex anterior or front edges of the lower canines to penetrate the animal's trachea or windpipe and major vessels, the paired jugular veins and carotid arteries. A backwards yank then severed these, ensuring that the prey would both asphyxiate and bleed out within minutes (figure 2.4). Compared to the sometimes arduous way an individual lion brings down its prey, risking a kick or goring from a large ungulate, the sabertooth's was a bloody but much faster way of killing.

The canines of early conical-toothed cats retained a more basal function of piercing or holding. In modern cats, when the upper and lower canines are forced past each other to pierce and grab the prey's body they "lock" into a powerful hold that can only be reversed by relaxing the jaws. The configuration of the cat's jaw joint only permits an orthal or "up and down" movement and doesn't allow the jaws to slip sideways and weaken the hold. The soft tissues penetrated by the conical canine tooth may exert a slight amount of counter-pressure and friction on the surface of the tooth, which would ordinarily make it harder to pull the canines out quickly and reposition them for another bite. To make extraction more rapid, felids and some other carnivorans have a groove known as a longitudinal canine sulcus on the labial (lateral or outer) and sometimes lingual (middle or inner) surface of the tooth, running parallel to the main axis of both the upper and lower canines. This minute air space alleviates the friction of the surface-to-tissue contact, thereby permitting a quick withdrawal of the canine.

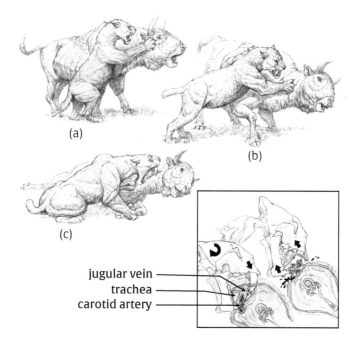

FIGURE 2.4. Machairodont shear bite.
Although sabertoothed cats such as *Smilodon fatalis* easily took smaller prey, tackling larger animals such as *Bison antiquus* probably required two or more cats (a, b). After the prey was immobilized (c) the cat positioned itself for a shear-bite to the neck (here seen in the boxed inset, at the cross section of the prey's axis vertebra) that would minimize the risk of breakage to its upper canines. Once positioned (skull at left), the canines made a shallow bite that sliced through and lacerated major blood vessels and the trachea (skull at right). An upward, jerking motion completed the damage, causing a bloody but quick death.

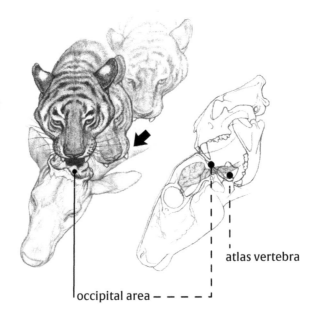

atlas vertebra

occipital area – – – –

FIGURE 2.5. Rapid test bites to locate intervertebral space.
When subduing a medium-sized prey animal like a deer, a tiger, aided by the
sensitivity of the gums along its upper canines, may make a series of "test bites"
along the upper neck to locate the space between the skull's occipital region and
the first (atlas) vertebra, the optimal area for the upper canines to penetrate the
brainstem and the jaws to clamp, enabling a quick kill.

Rapidly pulling out the canines is extremely important to a cat's
method of killing. Small cats, after grasping the prey with the forepaws,
deliver several rapid "test bites" to the prey's neck to locate the brainstem
or spinal cord through the gaps between the uppermost neck vertebrae
(figure 2.5). Once one of these gaps has been found, the final, killing
bite is administered. Lightning-fast speed is necessary not only to keep the
rodent or other victim from struggling free but also to keep it from biting
back; rodents have wickedly sharp incisors that can sever tendons and do
other damage to a predator. For smaller cats the canines function as long,
sharp daggers to quickly stab and penetrate. This use is aided by the fact
that cats and some other predators can "feel" with their canines, a situation
in which proprioceptor nerves supply a fine sense of location, pressure,
and movement. In felines both the upper and lower jaws have a large
space, or diastema, behind the canines. Evolutionary reduction and loss
of premolar teeth have separated the piercing and holding function of
the front teeth from the slicing function of the rear teeth, and this allows
the canines enough surrounding space to stab as deeply as possible.

To penetrate between the openings of its prey's cervical vertebrae a cat's canines must fit within the animal's intervertebral openings; for killing larger prey the canines should ideally be longer, broader, and more widely spaced than those used on smaller prey. To illustrate this, the zoologist David MacDonald (Oxford University) points to a study conducted by Tamar Dayan (Tel Aviv University) and her colleagues that shows a correlation among the canine sizes of three **sympatric** (same locality) small cats— the African or Near Eastern wildcat (*Felis sylvestris libyca*), the jungle cat (*F. chaus furax/prateri*), and the Asiatic caracal (*Caracal caracal schmitzi*)— that occur together both in Palestine and in the Sindh region of Pakistan. In both the Palestinian and Pakistani sympatric groups there was overlap in the sizes of carnassials regardless of species and gender (males are larger than females). This was *not* so in the canines. The length, diameter, and spacing from the incisors and premolars were distinct and separate by both gender and species. In the Pakistani sympatrics, although the size relationships were similar, they were "stepped down." This is because their habitat is shared by a *fourth* species, the fishing cat (*Prionailurus vivverinus*), whose canines, in accordance with its predation tendencies, are even larger.

This "search and stab" method of killing probably evolved early in felid history. Both cats and their less specialized, viverrid relatives, like modern genets, aim their bites toward the dorsal part of their prey's neck, just behind the skull. This "nape" of the neck is the widest, most optimal place for canine insertion. In the large-bodied pantherins, the nape-bite technique for killing small prey like a gazelle might involve crushing the first cervical or axis vertebra, and thereby the brainstem, instead of just piercing the intervertebral opening. However, the stronger, more muscular necks of larger prey can make a speedy nape bite too difficult. Instead, the jaws are used to suffocate the prey by clamping onto the underside of the neck to seal off the trachea or windpipe or by completely covering the nose, leading to asphyxiation. This can take time, sometimes as much as ten minutes.

The long whiskers, technically mystacial vibrissae, which end in nerve receptors and are embedded in the rounded, fleshy prominences on either side of the nose ("whisker beds"), are also vital in making a kill (figure 2.6). When the prey's neck is close enough for the "test bites," the point of contact of the canines is out of visual range. Muscles shift the normally sideways- and backward-directed vibrissae forward so that the ends make contact with the prey, supplying information that coordinates with the canines to accurately gauge the neck's position (figure 2.7). Small prey can be completely enveloped by these sensitive whisker tips, and the areas of the cerebral hemisphere on a felid brain that interpret touch from these and the preorbital and lateral vibrissae are well developed.

The sense of smell in felids was compromised by the need for a shorter muzzle, resulting in less space for the olfactory areas that are so well developed in dogs and bears. However, eyesight and hearing are acute, as one

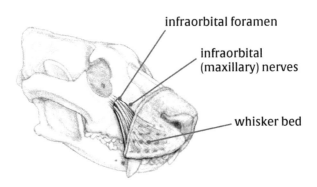

infraorbital foramen

infraorbital
(maxillary) nerves

whisker bed

FIGURE 2.6. Infraorbital nerve.
Skull of American lion, *Panthera atrox*, showing the infraorbital foramen (large opening beneath the orbit or eye socket) from which the infraorbital or maxillary nerves emerge. These supply the cat's whiskers with tactile ability, crucial in allowing it to position its jaws to bite (see figure 2.5).

FIGURE 2.7. Action of vibrissae.
A snow leopard with vibrissae or whiskers in a relaxed position (*left*) and with vibrissae moved forward (*right*), to sense the position of a bharal's throat when the cat is biting down.

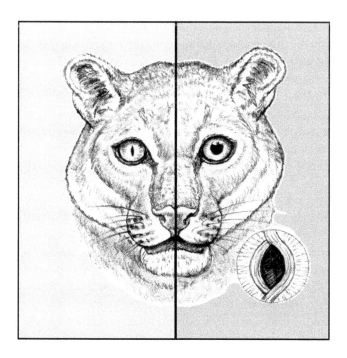

FIGURE 2.8. How a cat's pupils adapt to light.
Under bright light the eye pupil of a small cat species contracts to a slit (*left*) to reduce the incoming light rays to the correct amount needed by the retina to register an image; in larger cats the pupil assumes the shape of a circle rather than a slit. Under dim light the pupil expands (*right*) like the shutter of a camera because more light is needed. Pupil contraction is accomplished by retinal muscles that surround the pupil (*lower right*).

would expect in a highly nocturnal hunter. A cat's vision must be able to accommodate the much lower light levels of night as well as those of day. As with humans and other animals this is accomplished by having a pupil surrounded by a contractile iris that, like a camera's aperture leaves, adjusts its diameter in response to the amount of light entering. In bright, full light the pupil contracts to limit the amount of illumination to the necessary amount the retina needs to receive images, while in dim, minimal light the pupil expands to maximize the amount required for image perception (figure 2.8). Human pupils can close down considerably in bright light, but those of a small cat species can shrink to an extremely narrow slit and those of larger species contract to a small oval. Like other nocturnal animals, a cat's sensitive night vision is based on the many light receptor "rods" (noncolor photoreceptor cells, as opposed to the "cones," or color receptors) on the surface of the retina, but their light-gathering ability is also enhanced by a reflective layer of cells, the tapetum lucidum, located directly behind the retina. This reflects the incoming light back to the rods to be received a second time, ensuring that even more light will be collected and passed along to the brain to form an image. This reflecting layer is responsible for the red or green "glowing eyes" of animals we see at night when they are picked out by the headlights of a vehicle. The forward-facing eyes of a cat, like those of an owl or primate, provide excellent binocular vision for judging visual depth—another requirement for a hunter to calculate prey distances. Proportionately large orbits or eye sockets in felids correspond with nighttime visual acuity, and the relatively smaller orbits of some machairodontines have led paleontologists like Turner and Antón to suggest that these may have been largely diurnal or daytime predators. An exception among these may have been the homotherin sabertooths, which, in addition to larger eye sockets, had brains that show an enlarged visual cortex to aid in night hunting.

Little research has been conducted in testing the hearing ability of felids, but studies of domestic cats suggest potential auditory levels with frequencies higher than those of humans. It's reasonable to think that, as with vision, this also would be an essential sense for a largely nighttime predator. Field observations of lions show that they are able to hear well over long distances (as exemplified by the reaction of individuals to one that is roaring to proclaim territory) and can pick up sounds caused by the subtle vocalizations and movements of prey.

Recent studies of the extant, diurnally hunting cheetah *Acinonyx* by Camille Grohé (American Museum of Natural History) and colleagues reveal that this species possesses a distinctive vestibular canal system of the inner ear that is more highly developed than those of other carnivorans. The vestibular system, which provides a sense of spatial orientation in vertebrates, is vital in coordinating body movement and balance, and both the volume and shape of the vestibular canals are highly distinctive in *Acinonyx*, contributing to this cat's ability to execute sharp turns and keep its balance while pursuing fast-moving, typically antelopin prey. Such specialized structures are not found in other recent cats or in closely related fossil acinonychins (like *Acinonyx pardinensis*) that presumably had similar predatory habits.

Like many other mammals, felids have relatively large external ears (auricles or pinnae) that funnel auditory vibrations down to the middle ear and then to the auditory nerve. This in turn forwards stimuli and sends them to the brain. Bands of muscle underneath the skin can swivel the pinnae like radar dishes front, sideways, and back, and these instantly jerk to attention when prey is detected (figure 2.9).

Studies made from skull endocasts (the natural or artificial infillings of the cerebral cavity that preserve the general shape and details of the brain's outer surface) by Leonard B. Radinsky (New York Academy of Sciences) in the late 1960s, show that the cerebral portion of the felid brain underwent much evolutionary change in complexity (figure 2.10).

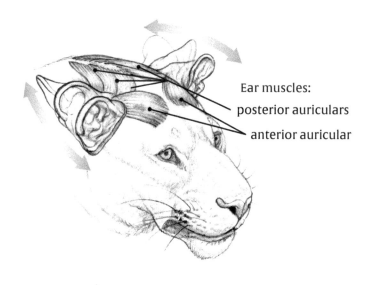

Ear muscles:

posterior auriculars

anterior auricular

FIGURE 2.9. Muscles that control ear position.
Although only the major ones are shown here, a felid has up to thirty-two auricular or ear muscles beneath the skin that can contract to pull the pinna, or outer ear, into a variety of positions from forward to backward, aiding the ear both to collect sound waves and to express emotions. Each ear can be moved independently of the other.

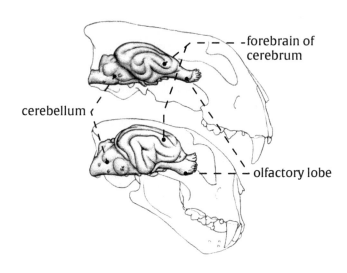

-forebrain of cerebrum

cerebellum

olfactory lobe

FIGURE 2.10. Evolution of felid brain.
Compared with the brain of a basal felid like *Pseudaelurus* (*above*), a modern cat like the puma (*below*) possesses a larger cerebellum (the structure that coordinates overall body movement), a larger olfactory lobe, and a larger, more complex cerebral forebrain. These correspond to the development of greater cognitive ability (in the case of the forebrain) and more sophisticated sensory and hunting abilities.

The cerebrum of the basal felid *Pseudaelurus* displays a relatively simple pattern of gyri (convolutions) and sulci (furrows or grooves), becoming more complex in *Proailurus* and ending with a much more developed, modern pattern of expansion and infolding in advanced machairodontines like *Smilodon*, small cats like *Felis*, and pantherins like *Panthera*. This is particularly noticeable in those areas dealing with hearing, vision, and sensorimotor processing. Similar developments occur from the early Miocene onward in the brains of the herbivore species that would have formed the cats' potential prey. The increased complexity and size of the mammalian cerebrum through time probably resulted in increased cognitive function. In all animals, neural "hard wiring" and instinctive responses are the basis of behavior, but adaptation to changing ecosystems probably required more complex behavioral responses and options both for predation and for avoiding predation.

A felid's acute senses, fast reflexes, and adaptive flexibility are what direct its skeletally strong, well-muscled body when capturing prey (figure 2.11). The earliest cats were arboreal hunters and already had supple spines and long, balancing tails to aid in leaping from branch to branch. These attributes have subsequently served most ambush predators and pursuit predators well with some notable exceptions, such as the advanced sabertooths, whose backs were less flexible and whose tails were short. When cats started to pursue terrestrial prey, however, the need for speed resulted in longer limbs for increased stride length and speed on the ground as well more reach in capturing. This was accomplished by slightly shortening the upper limb bones (humerus and femur) and lengthening the lower bones of the forelimbs (radius and ulna) and hindlimbs (tibia and fibula) as well as the **metapodials** of the feet (metacarpals below the wrists and metatarsals below the ankles [figure 2.12]). In climbing animals the metapodials and phalanges, or toe bones, are relatively short and widespread in order to maximize a stable, strong grip. The entire sole of the hindfoot comes into contact with a surface in what is known as a plantigrade stance, seen in a walking bear or raccoon. Running and springing, however, benefit from longer fore and hind paws that can flex and help thrust the body forward; for this reason the metapodials of cats and other fast-moving carnivorans became both longer and more closely bound together, giving them more strength. The hind foot in particular is a powerful lever for creating forward thrust, and the earlier plantigrade stance was abandoned for a digitigrade one, in which the sole is off the ground and only the toes make contact (figure 2.13). The ankle joint, too, changed from one that permitted much lateral rotation, good for climbing trees, into one that was limited to fore-and-aft motion, stronger and better for the stresses of fast forward movement. Some smaller cats like ocelots never completely abandoned their ability to remain effective climbers, however, and the clouded leopard retains an ankle joint that allows it to hang completely from a tree trunk with its hind feet pointed almost backward. Clouded leopards as climbers have also reversed the trend toward longer lower limbs, secondarily evolving shorter elements because this gives the limb proportions a better mechanical advantage when moving in trees.

Relatively long thoracic and lumbar spinal vertebrae are also an asset to a terrestrial runner. This is because the overall flexibility of the vertebral series allows the back to be doubled up and extended like a spring, working with the limbs to achieve a powerful forward thrust that increases stride. As we learned from the problem of stamina, however (see chapter 1), an animal's potential to maximally evolve in one area may be limited or compromised by the need to maintain or develop other factors affecting its capabilities. In the case of cats, the benefits of a long back inherited from tree-dwelling ancestors had to be balanced by the need for strong skeletal support as cats grew larger. As shown by Turner and Antón in their 1997 book, many nimravid and machairodontine sabertoothed cats actually evolved shorter, less flexible backs; although this reduced the ability to run when ambushing prey, a shorter back conferred a better mechanical advantage and greater stability when grappling with prey from an upright position. At the opposite extreme are cheetahs, whose long, flexible backs combine with limb length in giving them the longest strides for their size, greatest temporary top speed, and fastest acceleration of any land animal. Subtle differences in skeletal proportions give anatomists and paleontologists a reasonably accurate idea of the different locomotory functions and, as result, the probable hunting procedures of ancient felids (figure 2.11b). As we'll learn in chapter 3, this was crucial to the evolutionary success of the pantherin cats.

Although cats are not unique among carnivorans in having retractable claws (a few members of the living viverrids have them, as did extinct nimravids and creodonts), these are especially well developed in cats and important for capturing prey. Claws are made of **keratin**, the same material as a human finger- or toenail, and continuously grow out of a recessed area in the front of the terminal (or distal) phalanx (toe bone). Both the claw and the club-shaped phalanx are normally held in a retracted position by elastic ligaments up and back along the lateral side of the second phalanx, but when the cat needs them they are extended down and forward by tendons and muscles to become effective grappling hooks (figure 2.11c). Mounted on each of the splayed-out forepaw's five toes, a cat's extended claws are effective at hooking into and pulling prey close enough for the jaws to bite; wielded at the end of the muscular forelimbs they are weapons for creating deeply wounding lacerations. The dew claw of the first digit, as in canids, is usually short, but in the sabertooths, it's unusually large, hooklike, and sharp, with the short length and its mechanical advantage conferring great strength. This is also true of the cheetah, whose claws are not fully retractable but include an especially enlarged dew claw of the forepaw, which it uses to trip or slow down its prey.

FIGURE 2.11. (a) Pantherin skeleton and musculature. Although not capable of sustained speed, the body of a pantherin like the tiger combines both suppleness and great strength. An adult big cat can make sudden, explosive charges to overtake prey, and its powerful limb, neck, and jaw muscles are usually more than enough to overcome large animals. While the prey is kept from escaping by formidable claws, the canine teeth make the kill either by piercing the base of the neck, lacerating the throat, or causing suffocation by clamping over the nose.

(b) Differences in felid spinal anatomy.

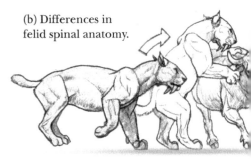

The shortened spines of sabertooths produced enhanced stability for the cat while they wrestled prey to the ground.

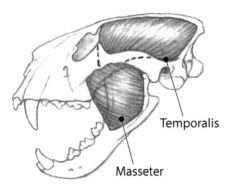

Temporalis

Masseter

The powerful bite of a big cat like the leopard shown here is created by two major, paired muscles, the masseter of the cheeks and the temporalis at the sides of the head. A leopard's bite force is 800 pounds per square inch and a tiger's is 1,000.

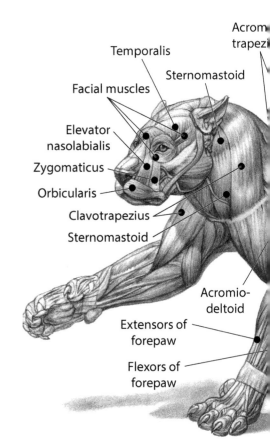

Acrom
trapezi

Temporalis

Sternomastoid

Facial muscles

Elevator
nasolabialis

Zygomaticus

Orbicularis

Clavotrapezius

Sternomastoid

Acromio-
deltoid

Extensors of
forepaw

Flexors of
forepaw

(c) Retraction of claws.
To keep their claws sharp, almost all cats are able to completely retract the third, terminal and claw-bearing phalanx of each toe away from the ground and alongside the second (gray arrow, *top*), by contracting a lower muscle, relayed by a tendon, in the toe (black arrow, *top*). To extend the claw, an upper muscle contracts (black arrow, *bottom*) through a tendon, pulling the phalanx forward (gray arrow, *bottom*).

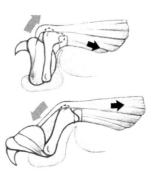

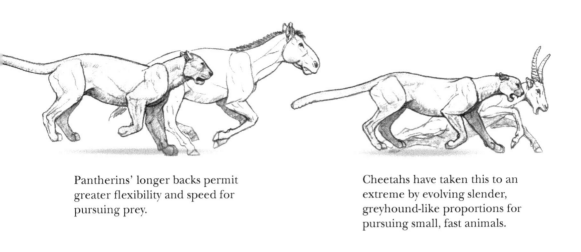

Pantherins' longer backs permit greater flexibility and speed for pursuing prey.

Cheetahs have taken this to an extreme by evolving slender, greyhound-like proportions for pursuing small, fast animals.

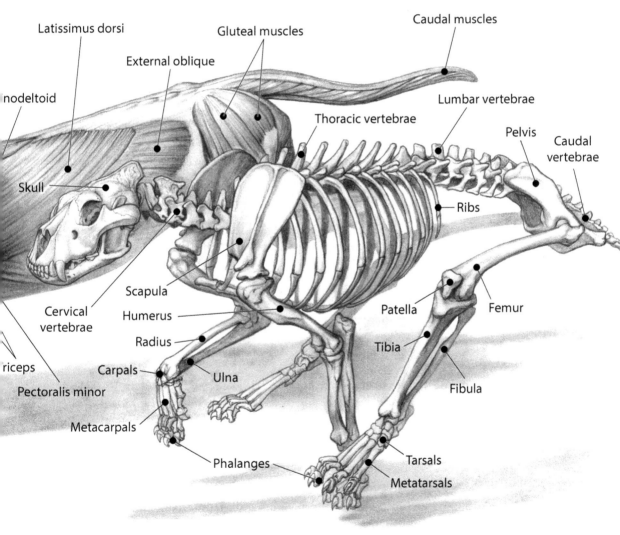

Latissimus dorsi

Gluteal muscles

Caudal muscles

External oblique

nodeltoid

Lumbar vertebrae

Thoracic vertebrae

Pelvis

Caudal vertebrae

Skull

Ribs

Scapula

Cervical vertebrae

Humerus

Patella

Femur

Radius

Tibia

riceps

Carpals

Ulna

Fibula

Pectoralis minor

Metacarpals

Phalanges

Tarsals

Metatarsals

Much of a big cat's strength lies in the effective muscular attachments of the limbs and back. The powerful flexor and extensor muscles are anchored to prominent ridges, crests, and bony expansions, keeping skeletal mass to a minimum while maximizing areas where the muscles originate and insert.

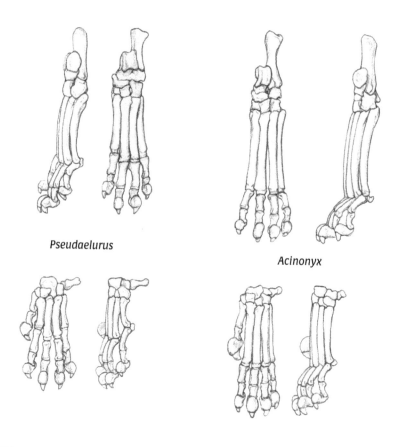

Pseudaelurus

Acinonyx

FIGURE 2.12. Modification of metapodials in early and later cats.
A comparison of the hindfoot metatarsals (*top row*) and forefoot metacarpals
(*bottom row*) from the skeletons of the Early Miocene basal felid *Pseudaelurus
validus* and the specialized, extant cursorial cheetah *Acinonyx jubatus* show the
lengthening of these elements. This, as well as the lengthening of the tibia/fibula
(lower leg) and the shortening of the femur (upper leg), contributed over time to
greater stride length and greater speed in running.

The coat color and patterning of living cats has a direct relationship
with their behaviors and is highly important in ensuring their success as
predators as its main purpose is concealment to minimize the chances of
detection by prey. The coat pattern of extinct species is necessarily conjectural because hair is rarely fossilized and because few likenesses of extinct
cats have been recorded in cave paintings. A pattern of large, dark, irregular splotches or streaks may be the most basal since it would have allowed
the animal's body to blend with the light and shadow patterns of primeval
forest vegetation. This concealing or cryptic design is seen in the clouded
leopard, which often hunts in tree-canopy habitats that are broken up by
large areas of light and shadow. When felids became more terrestrial, this

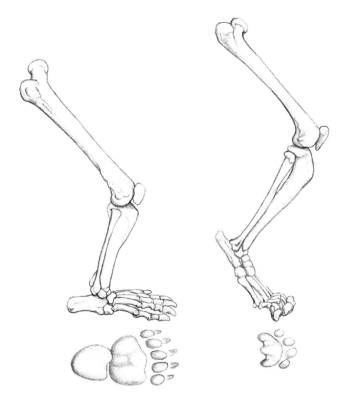

FIGURE 2.13. Plantigrade and digitigrade stance.
Although a plantigrade stance, in which the foot contacts the ground from toes to heel, is effective in providing support for a large animal like a bear (*left*), a digitigrade stance, in which the foot makes ground contact only at the toes, is typical of fast-moving forms (lion, *right*). This increases the amount of muscular leverage and, hence, the amount of thrust and speed produced by the foot.

probably changed into a more complicated disruptive design of elongated streaks and spots, calling attention away from the body contours to be effective in a variety of environments ranging from closed or open woodlands to savannas, and could have resembled the coat of an African serval (*Leptailurus serval*). In such mixed environments a system of open rosettes and streaks may possibly be more effective for a larger-bodied species like a leopard, and in forest-grassland margins open and double vertical stripes, like a tiger's, seem to produce the best concealment (figure 2.14). This hypothesis is conjectural but offers a plausible path for the kind of patterning diversity seen in felids. In some of the large pantherins, definite variations in patterns (and color) occur that are typical of subspecies in different regions and may relate to their value for concealment. For example, the lighter background color and fainter stripes of the Amur

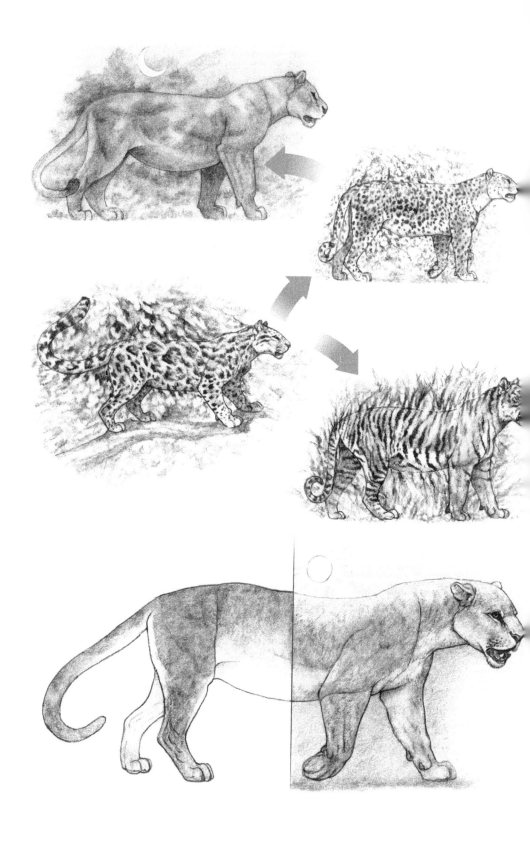

tiger are probably better adapted in an environment of seasonal extremes than are the coppery-colored coat and darker, more contrasting stripes of Indian tigers in subtropical habitats. It's also possible that similar patterning plays a part in the group recognition of individuals in a particular subspecies or region. The question then arises, "What about more or less monochrome-coated species like lions, caracals, and others?" The answer is that a monochromatic coat can be surprisingly hard to detect when an animal stands still against some vegetation backgrounds, but lion cubs have spots when newborn, and the spotting persists on the bellies of some juvenile individuals. In temperate environments hormonal changes produce a transition in North American pumas, and their deer prey, from reddish-tan coats to grayish ones when fall and winter arrive, in keeping with the drab colors and more low-angled lighting of those seasons. In less seasonal tropical and subtropical climates some monochromatic cats' coats vary according to region; the coats of Asian lions (*P. leo persica*) are noticeably grayer all year long than those of their tawny-beige African cousins. In the far north, Pleistocene steppe lions (*Panthera spelaea*) and homotherin sabertooths may have become mostly or totally white as winter approached, like the extant Arctic, white, or polar wolf (*Canis lupus arctos*), effectively concealing themselves from prey—as shown by Antón's 1997 depiction of a *Homotherium serum* that has just brought down a similarly colored dall sheep. We may never know for certain, but for this and other extinct cats it's fascinating to conjecture.

FIGURE 2.14. Possible fur pattern evolution in cats.
From a large pattern of splotches that offers disguise in a humid tropical forest's light and shadow (clouded leopard, *center left*) a finer pattern more effective in drier temperate forest may have evolved (leopard, *upper right*) while stripes became optimal in grasslands (tiger, *lower right*). Even monochromatic coats like those of lions (*upper left*) offer effective concealment in open settings from dusk or dawn, when these cats prefer to hunt. Most animals have coat coloration that is darker on top than on the bottom (female, *below, left half*). When this is lit from above (*right half*), the dark top looks lighter while the light underbelly looks dark, equalizing both and making the animal harder to see. This principle, called countershading, works equally well for both prey and predators, but big cats also have patterns that help conceal them from prey by breaking up their body contours.

CHAPTER 3
A Breath of Frost

Zanda Basin, Tibet, 4.4 million years ago: The patch of bog sedge high up on the cliff is less accessible than those lower down but thick and attractive in the feeble, cold morning light. It had escaped the notice of the small herd of ancestral bharal as they foraged for grasses along the cliff face but now three ewes, one followed closely by a lamb, scramble up to the fissure where the sedge is growing. As they feed, they are patiently watched by a lynx-sized, gray-blotched cat that has hidden itself behind boulders several yards away. At the moment that all heads are down he bounds forward, aiming for the lamb, which saves itself by taking its mother's cue and jumping sideways to the ledge below. The cat, *Panthera blytheae*, flexes its back once more and makes a spectacular leap downward in pursuit, spraying rocks as it attempts to hook the lamb's left hind leg with its foreclaws. It misses only by inches but the lamb, in full flight along with the rest of the herd, escapes down the cliff and is gone, leaving the cat with nothing but the fading echoes of hoofbeats (figure 3.1).

. . . . .

The late Miocene mountains that surrounded the Zanda Basin of southwest Tibet saw many such failed hunts, and many other successful ones, as the land rose by several centimeters year by year, a result of the uneven and sometimes rapid uplift of the huge area that would one day become the Tibetan Plateau. This rise began after the colossal raft of the Indian subcontinent finally collided with the Eurasian tectonic plate to create the Himalayan Mountains about 50 million years ago. There is general agreement that the elevation of the Himalayas and Tibetan Plateau began between 60–50 Ma, with significant increases in elevation occurring the early Miocene (21–17 Ma), late Miocene (11–8 Ma), and, most recently, within the last 3 million years. Ringed by the rim of the High Himalayas at its southern border and continuing north into Mongolia and China as a system of generally northwest-to-southeast oriented, extensive mountain chains and basins, the region's huge size and height today profoundly affect the climates of the rest of Asia (figure 3.2).

During the later Miocene a major cooling trend, fueled by a polar temperature drop and east Antarctic ice growth, created a profound effect on overall worldwide ecosystems. In Eurasia this cooling resulted in

---

FIGURE 3.1. *Panthera blytheae* ambushing bharal herd in Tibet.

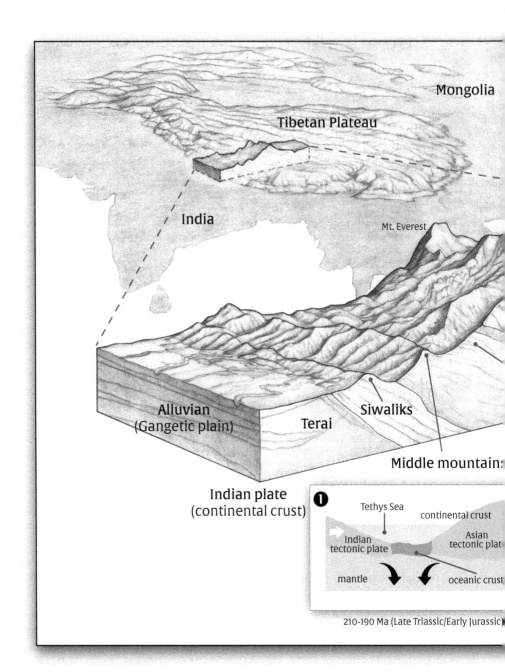

FIGURE 3.2. Rise of the Tibetan Plateau.
This series of diagrams shows in simplified form the development of the region.
(1) By the Late Triassic/Early Jurassic the Indian tectonic plate had drifted far
north (white arrow), where convection currents caused the oceanic crust to begin
buckling and subducting under the Asian plate beneath the Tethys Sea. (2) As
subduction proceeded, continued pressure caused the elevation of the Asian
plate (white arrow). At the same time rivers deposited sediments (black dots) that
gradually accumulated on the floor of Tethys. (3) During the Late Cretaceous/
Early Tertiary the accumulation and uplift led to the eventual disappearance of
Tethys, and the former sediments under pressure became rocks, which were thrust

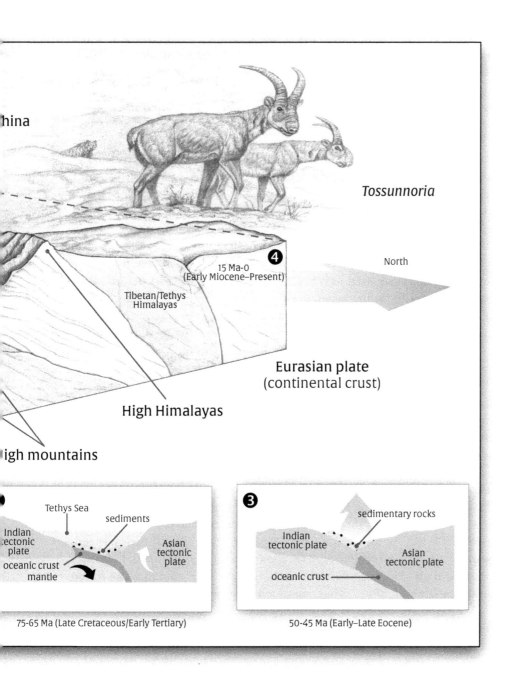

hina

Tossunnoria

4 North

15 Ma-0
(Early Miocene–Present)

Tibetan/Tethys
Himalayas

Eurasian plate
(continental crust)

High Himalayas

igh mountains

Tethys Sea

sediments

Indian
tectonic
plate

Asian
tectonic
plate

oceanic crust
mantle

75-65 Ma (Late Cretaceous/Early Tertiary)

3

sedimentary rocks

Indian
tectonic plate

Asian
tectonic plate

oceanic crust

50-45 Ma (Early–Late Eocene)

upward (arrows). (4) From the Early Miocene and continuing until the present, first the Tibetan/Tethys Himalayas were formed, followed by the High Himalayas, the high mountains, the middle mountains, and, finally, the Siwaliks (Mahabharats). The Terai strata farther south (in large drawing, at left) belong to the continental crust overlying the Indian plate and are covered by alluvial deposits of the Gangetic Plain. Basal caprins such as *Tossunnoria* (*above right*) adapted to the increasingly cold and dry conditions of the Tibetan Plateau and provided prey for the evolving pantherins. *Source*: Diagrams after Mishra and Jeffries (1991).

a progressive shift away from closed forest to open woodlands and large tracts of savanna, or grassland-dominated environments. The very latest Miocene saw a temporary return to the warmer climatic conditions in northern China and elsewhere in Asia that favored humid, laurellaceous evergreen forests, but the long-established warm, rain-laden winds from the southeast began to be blocked as the Tibetan Plateau grew higher. This in turn produced a shift toward cooler winters and decreased summer rainfall and a much cooler, more arid grass-dominated **steppe**, or treeless prairie environment, in the Tibetan Plateau region. Here and in other regions this predominance of grass favored the natural selection of mammalian herbivores with hypsodont or high-crowned molar teeth to deal with this comparatively low-calorie but abundant food source. Grass contains tall spicules of the mineral silica in its cell walls to help it stay upright, so chewing grass is like chewing sandpaper. The greater height of hypsodont molars, compared to a brachyodont or low-crowned dentition for chewing leaves and softer vegetation, was an adaption to make an animal's permanent teeth last longer throughout its lifetime when consuming this more abrasive vegetation or other grit-covered food (figure 3.3).

These worldwide climate changes not only favored the spread of grasslands but also affected their composition. Until the end of the Miocene, climatic conditions had favored the CAM photosynthetic process for succulent plants and the C_3 photosynthetic process for all other plants. However, near the epoch's end (about 7 Ma) a continuing decrease in atmospheric CO_2 favored a different photosynthetic pathway, C_4, which was adopted by grasses (and a few shrubs) that grew at lower elevations in temperate to tropical areas with summer rains. These plants formed the diet of ungulates such as horses and rhinos that lived on what is now the Tibetan Plateau at the end of the Miocene, indicating that the Tibetan Plateau was lower at that time and the climate there was warmer and wetter than it is today. With the continuation of the Tibetan Plateau's uplift this changed; as the elevation approached 3,000 meters and the region cooled, C_3 grasses replaced their C_4 predecessors. Some herbivores were able to adapt to the locally cooler conditions; others migrated elsewhere or became extinct.

We know this thanks to yet another new investigative technique called stable isotope analysis (figure 3.4). Isotopes are naturally occurring different variants of specific elements. The different isotopes of an element have the same number of protons in their nucleus but have different numbers of neutrons. The common element carbon, for example, has four isotopes, two of which—C_{12} and C_{13}—are stable and the other two—C_{11} and C_{14}—are unstable. An unstable isotope disintegrates through time, and its rate of disintegration is measured by its half life—the amount of time needed for its concentration to decrease by half. The half life of C_{11} is twenty minutes, so this isotope can't be detected in fossils. The half life of C_{14} is 5,700 years, so half of it will have disintegrated in 5,700 years after the animal's

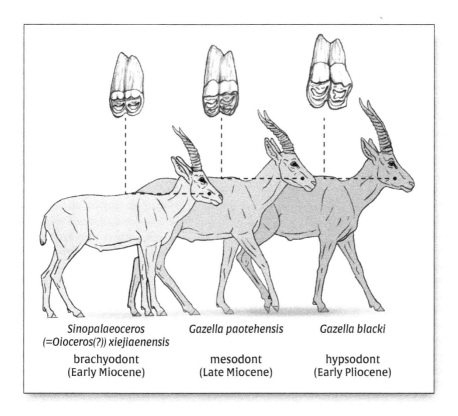

Sinopalaeoceros
(=Oioceros(?)) xiejiaenensis
brachyodont
(Early Miocene)

Gazella paotehensis
mesodont
(Late Miocene)

Gazella blacki
hypsodont
(Early Pliocene)

FIGURE 3.3. Changes in brachydont to hypsodont dentition.
As the Tibetan Plateau and surrounding areas grew higher, cooler, and drier during
the latest Miocene, grasses became a more dominant food source for bovids and
some other ungulates. Although as in other bovids the teeth of gazelles retained a
conservative selenodont enamel-and-dentine occlusal surface, fossil gazelles from the
Linxia Basin of northwestern China show an evolutionary trend from low-crowned
(brachyodont) teeth, primarily suited for processing softer shrubs and forbs and
found in *Sinopaleoceros (=Oioceros?) xiejiaenensis,* through the intermediate (mesodont)
condition of *Gazella paotehensis* to the high crowned (hypsodont) teeth of *Gazella blacki*
that could better withstand the wear from abrasive grasses and soil grit.

death, three-quarters will have disintegrated in 11,400 years, seven-eighths
in 17,100 years, and so on. After about 50,000 years the proportion of C_{14}
left in the tissues is too small to be measurable. Thus, C_{14} provides a useful
way of directly measuring the age of fossils that are less than 50,000 years
old. The age of older fossils has to be estimated using other methods.

Carbon is a major constituent of plants, so this element becomes
absorbed into the bodies of plant-eating animals and into those of the
carnivorans that eat them. Preserved body parts (teeth, bones, horns, hair,
etc.) can then be chemically analyzed to determine the relative concentrations of isotopes. The stable isotope C_{13} is less common than C_{12} but is

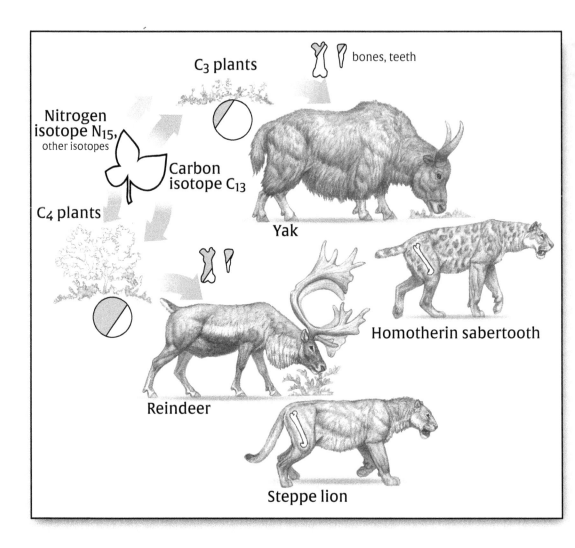

FIGURE 3.4. Stable carbon isotope analysis.
Plant species each have distinctive ratios of stable carbon isotopes (C_{12}, C_{13}) that reflect their photosynthetic pathway (C_3 or C_4). When plants are eaten by herbivores, these isotopes are absorbed into the body tissues and their proportions in relation to one another stay the same. Chemical analyses of fossil bones, teeth, hair, or horn can indicate if a given herbivore species ate mainly trees and shrubs (typically C_3) or mainly grasses (C_4 in areas with summer rain, C_3 in high, cool, and dry regions) and by implication whether it lived in a woodland, a closed forest habitat, or an open savanna. When the fossil tissues of a carnivore are analyzed, these same isotope proportions in turn indicate from which habitat their prey was derived. Analysis shows that during the coldest, most recent glacial phase, homotherin sabertooths preyed heavily on yak in Asia whereas steppe lions relied mostly on reindeer in Europe and bison in Asia.

important for environmental analysis because it occurs in different concentrations in C_3 and C_4 plants. Each isotope has a characteristic digital "signature" when measured by a device called a mass spectrometer. In the case of carbon, the proportion of C_{12} to C_{13} in the body tissues may indicate if an herbivore ate mostly C_3 or C_4 plants. Thus the proportion of C_{13} in an herbivore's tissues may indicate whether it browsed on leaves or grazed on grasses or whether it was feeding in an open habitat or closed forest. In a sequence of dated fossils, the carbon signatures of an herbivore's bones and teeth can be used to track the dietary trends of the animal's lineage through time, whereas those of a carnivoran provides essential information about the animals on which it was feeding (see chapter 6).

The late Miocene–earliest Pliocene resident mammals of the Tibetan Plateau's Gyirong Basin were part of a faunal community known as the Chinese Hipparion fauna (figure 3.5). This was characterized by browsing, C_4-adapted forms that indicate a warm, temperate forest-grassland environment that still predominated during the late Miocene. Potential prey animals included pikas (*Ochotona guizhongensis*), a squirrel (*Aepyosciurus* sp.), a rabbit (*Trischizolagus mirificus*), a basal short-necked giraffe (*Paleotragus microdon*), a deer (*Metacervulus capriolinus*), a gazelle (*Gazella gaudryi*), a camel (*Paracamelus gigas*), a hipparionine horse (*Hipparion guizhongensis*), and a browsing rhino (*Chilotherium xizangensis*). When the Tibetan Plateau's climate and vegetation started to shift, some species stayed and adapted, whereas others migrated to other areas or became extinct.

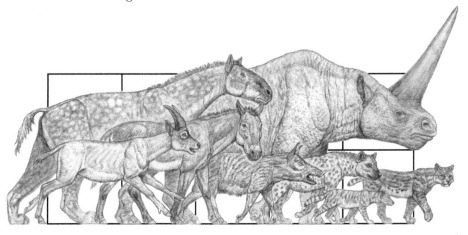

FIGURE 3.5. Chinese Hipparion fauna.
Assorted species typical of the Late Miocene Chinese Hipparion fauna of the Baodean Asian LMS, 7.3–6.4 Ma. From left to right, the chalicothere *Ancylotherium pentelicum*, the ovibovin *Urmiatherium intermedium*, the horse *Hipparion zandaense*, the pig *Chleuastochoerus stehlini*, the elasmotherin rhino *Sinotherium lagrelii*, the bone-crushing hyena *Adcrocuta exemia*, the ictitherid hyena *Ictitherium wongi*, and the metailurin cat *Metailurus major*. The side of each square equals 1 meter (3.28 ft.).

Among the artiodactyls one family in particular, the **Bovidae**, experienced an explosive radiation during the mid- to late Miocene. The bovids comprised a diverse group of ruminants (foregut digesters) with unbranched, bony horn cores covered by a keratinous sheath and include antelopes, gazelles, cattle, goats, sheep, and their relatives. From the ancestral bovids evolved the Eurasian-based tribes **Saigini**, represented today by the living chiru or Tibetan antelope (*Pantholops hodgesoni*) and saiga antelope (*Saiga tatarica*); **Ovibovini**—including the takin (*Budorcas unicolor*) and muskox (*Ovibos moschatus*); **Caprini**—made up of various species of goats (*Capra* spp.) and sheep (*Ovis* spp.); and **Rupicaprini**, which today include the serows (*Capricornis* spp.) and gorals (*Nemhoraedus* spp., figure 3.6). Although the fossil record is sparse, genetic evidence

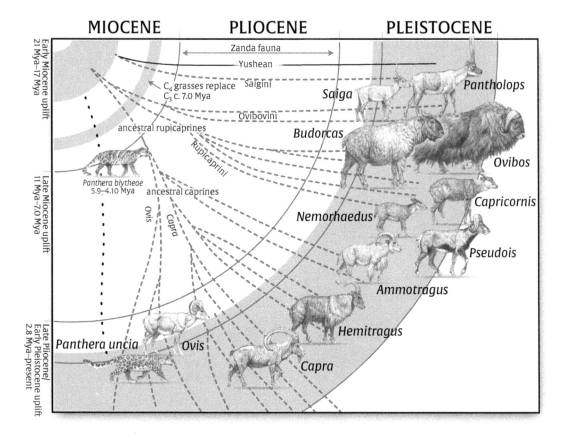

FIGURE 3.6. Evolutionary radiation of the Tibetan region saigins, ovibovins, rupicaprins, and caprins.
Duration of the Pliocene is proportionately exaggerated in relation to other epochs in order to clearly show estimated clade divergences and key events, such as stages of uplift (animals to scale). *Source:* Modified from Schaller (1977).

indicates that all these groups (except the Saigini) are closely related, originating about 11 Ma. Beginning as relatively small, unspecialized ungulates with compact bodies and simple, backward-directed, dagger-like horns, its members flourished during the late Miocene and increased in body size. Except for sheep, the caprins underwent a key adaptation in which the length of their metacarpals (the bones connecting the wrist to the fingers) was reduced. This shortening of the metacarpals gave better leverage in climbing and enabled caprins to exploit mountainous ecological niches not occupied by other ungulates.

With the rise and continued cooling of the Tibetan Plateau region, coolness- and aridity-adapted C_3 grasses and shrubs replaced the warmth- and moisture-requiring C_4 types. Those herbivores that stayed in the region variously developed insulating fur undercoats, nasal modifications, sophisticated thermal physiologies, and, in some cases, the ability to scamper and jump in the steep, mountainous terrains created by the continuing uplift (figure 3.7). The earliest forms of the rupicaprins, represented today by the

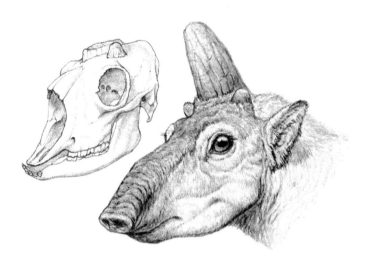

FIGURE 3.7. *Tsaidamotherium hedini,* skull and life restoration.
During the Late Miocene, rupicaprins and caprins on the Tibetan Plateau became specialized for living in high, cold altitudes, and one ovibovin, *Tsaidamotherium,* developed a wide, vaulted muzzle, similar to that of the living saiga antelope, to help warm inhaled air it before it entered the lungs. The two known species discovered in the Linxia and Qaidam Basins are unusual in that both had bizarrely mismatched horns: in one, *T. brevirostrum,* the right horn was short and flattened, whereas in *T. hedini* this horn was tall, conical, and much larger than the one on the left. Ruminant cranial appendages are notoriously variable so, without a larger sample, we can't tell if the bilateral asymmetries of these horns define their species, represent some kind of sexual dimorphism, or are merely examples of individual variation.

serows and gorals, were short-legged and stocky in build. As the area became increasingly mountainous, these attributes would have predisposed them to succeed. The ancestral rupicaprins began to adapt to the Tibetan Plateau's coldness and aridity during the early Pliocene about 5.10 Ma, and some species would eventually become **endemic**, that is, restricted to this region. In contrast, two tribes—the Saigini (chiru and saiga antelope lineages) and Ovibovini (takin and muskox)—had actually begun to evolve in the region far earlier, following the first pulse of uplift during the early Miocene. Instead of mountainous areas, these bovids preferred the similarly dry and cold but less hilly and more gently rolling basins, where extreme seasonal temperature fluctuations affected grass availability and made migratory behavior a dominant factor in their lives. By the early Pliocene they had been joined in the basins by some species of sheep, whereas immigrant goats and rupicaprins exploited steeper habitats.

All of these bovids responded to the region's low temperatures by evolving a dense layer of fleecy fur that underlay the coarse outer hairs of their coats, protecting against both cold and moisture. A corresponding adaptation was an increase in body size. Among many land-dwelling clades of endotherms (specifically, "warm blooded" mammals and birds) that have a broad geographical distribution, there is a tendency for species or subspecies inhabiting colder climates to be larger than ones in warmer areas. Known as **Bergmann's rule**, this is not a strict, invariable law but often applies to animals inhabiting a range of latitudes; Amur tigers and leopards of northern Asia are, for example, larger than Indian and southeast Asian tigers and leopards, with Amur tigers sometimes reaching up to 320 kg (700 lb.), while southeast Asian tigers typically range from 80–120 kg (176–264 lb.). It can also apply to altitude, where species in higher, colder elevations are larger than closely related species from lower, warmer elevations. This is the case with both the serows and gorals, in which Himalayan representatives like the high-altitude Himalayan serow (*Capricornis thar*) and Himalayan gray goral (*Naemorhedus bedfordi*) are larger, longer in body length, and taller at the shoulder than their lower-elevation equivalents, the Formosan serow (*Capricornis swinhoei*) and red goral (*Naemorhedus baileyi*) (figure 3.8). The explanation for Bergmann's rule is that animals of greater body size have a smaller surface area (where heat is collectively gained or lost) in proportion to their internal volumes and that as this proportion increases the body becomes more efficient in retaining warmth—a distinct advantage for a cold-climate species. Larger size has other benefits as well: to sustain its metabolism a bigger animal requires relatively fewer calories to move around and keep warm than a smaller one, calories that can mean the difference between life and death in a land of frigid winters and sparse resources.

However, Bergmann's rule doesn't explain all instances of exceptionally large size in some animal populations or subspecies. Although, as

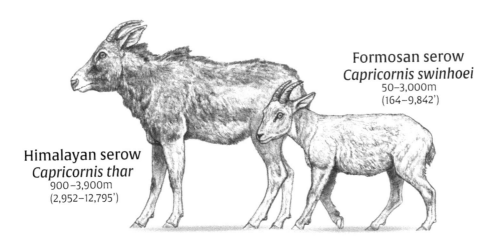

Formosan serow
Capricornis swinhoei
50–3,000m
(164–9,842')

Himalayan serow
Capricornis thar
900–3,900m
(2,952–12,795')

Himalayan gray ghoral
Naemorhedus bedfordi
900–2,750m
(2,952–9,022')

Red ghoral (India, Tibet, Burma)
Naemorhedus baileyi
1,000–2,000m
(3,280–6,561')

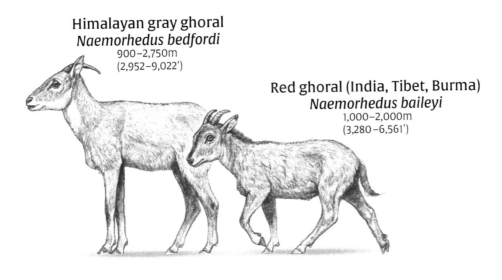

FIGURE 3.8. Bergmann's rule in relation to species and the altitude of their habitat ranges.

Some Asian rupicaprins such as serows and ghorals, in spite of their generally shared preference for lightly forested, hilly habitats, show size ranges corresponding to the altitudes at which they are found. This is a corollary of Bergmann's rule, the informal principle holding that larger size confers an advantage for a mammal's heat retention, which is adaptive for species living in colder surroundings. Larger, mountain-living rupicaprin and caprin species provided a challenge for the Tibetan Plateau's early pantherins, which subsequently increased in size to be able to prey on them.

already pointed out, living southeast Asian tigers are smaller than their northern counterparts, one of the largest of all known tigers occurred in the late Pleistocene on Java. Known as the "Ngandong tiger" (*Panthera tigris soloensis*) after the location in which they were found, some individuals reached more than 300 kg (661 lb.)—rivaling weights of the largest known living Amur (Siberian) tigers. In a 2015 study, Rebekka Volmer (University of Athens) suggested that the larger body mass in the Ngandong tigers was an adaptation to minimize competition with two of the region's other main sympatric predators, the large Merriam's dog (*Megacyon merriami*) and the dirk-toothed *Hemimachairodus zwierzyckii*—one of the region's two machairodonts—through reducing niche overlap in the three species' preferred prey sizes. During the earlier early to middle Pleistocene, the tigers and dirk-tooths were of similar size and were competing with the smaller but pack-hunting Merriam's dog. The larger size of the Ngandong tiger would have given it a selective advantage over the other two. At the end of the Pleistocene the dirk-tooths became extinct; Merriam's dog was replaced by the smaller, jackal-like, and now extinct "Trinil dog" (*Mececyon trinilensis*); and the Javan tigers reverted to a smaller size.

The large size of the Ngandong tiger has an interesting parallel in the mid- to late Pleistocene of Eurasia, where the large size of the steppe lions—*Panthera fossilis* and later *Panthera spelaea*—may have been a response to competition with large sympatric predators such as the giant, short-faced hyena (*Pachycrocuta brevirostris*), the spotted "cave hyena" (*Crocuta crocuta spelaea*), the homothere (*Homotherium latidens*), and the pack-hunting gray wolf, (*Canis lupus*). The appearance of the recently discovered giant Kenyan lion in East Africa (see chapter 4) may have a similar explanation.

The bovid communities, meanwhile, provided an opportunity for any carnivore that could bring down an agile, mountain-dwelling prey animal of moderate size. The best candidates for this within the Late Miocene–Early Pliocene Tibetan carnivoran community were the felids, whose relatively long, springlike backs; strong, clawed forelimbs; and powerful bite equipped them for killing swift and often vertically mobile prey. Although some medium to large sabertooths of the time, such as *Machairodus nihowanensis* and *Homotherium hengduanshanense*, had ranges that extended into the Tibetan Plateau, these cats were best able to ambush prey after a short rush from a concealed, forest-margin location on relatively level terrain. Because of their generally shorter, less flexible spines, machairodontines were not well adapted for leaping after small, agile prey that could escape by running up and down steep hillsides. Thus the rising Tibetan Plateau presented the feline cats with an open opportunity.

In the late summer of 2010 a field team led by the mammalian paleontologist Xiaoming Wang (Natural History Museum of Los Angeles County) and his associates was methodically exploring a section of Zanda Canyon in the southwest Ngari District of the Tibet Autonomous Region when one member, Juan Liu (State University of New York, Buffalo), came across some small, brown bones weathering out of a lens of greenish, coarse-grained sandstone (plate 2a, b). The bone bed was within a stratum dating to about 4.42 Ma, in the very late Miocene / early Pliocene Zanda Formation, and the bones turned out to be the cranium and partial mandible of a new pantherin cat, **Panthera blytheae**. According to Jack Tseng (State University of New York, Buffalo), Wang, and colleagues, the battered skull, although dorsoventrally crushed, is complete enough to show that this cat is definitely of the pantherin grade because of the arrangement of the upper skull sutures and other features (figure 3.9). About 20 percent smaller than that of the ounce or snow leopard (**Panthera uncia**), *P. blytheae* shares with it characters such as canines with an almost circular cross-section; a moderately inclined mandibular symphysis; a smooth transition between the symphysis and mandibular rami; a frontonasal depression or pit;

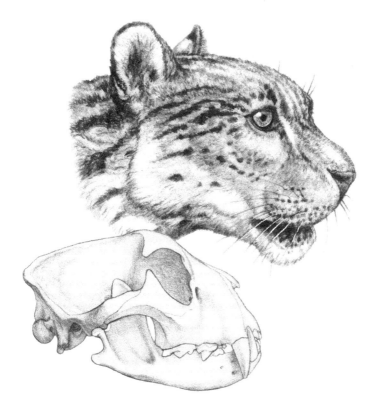

FIGURE 3.9. Skull, restoration of *Panthera blytheae* (see also plate 3).
About the size of a modern lynx or serval, the early pantherin *Panthera blytheae* may have resembled the smaller living Andean mountain cat or guigna (*Leopardus jacobita*) in its habits and was an agile hunter probably capable of taking a wide range of prey, including the ancestors of the modern bharal or blue sheep. Its dentition and cranial sutures indicate that it is the earliest known pantherin and the possible direct ancestor of the modern snow leopard.

a narrow distance between the anterior edge of the tympanic bullae and the glenoid ridge; a sharply turned ventral premaxillary-maxillary border at the canines; and, finally, a straight and symmetrical upper fourth premolar (P4) cusp alignment. *P. blythae*, however, is distinguished from *P. uncia* (and all other *Panthera* species) by the small labial cusp on the posterior cingulum of the upper third premolar (P3) and converging ridges on the labial surface of P4. Like *P. uncia*, *P. blytheae* has broad frontal bones, characteristic of modern pantherins, which suggests relatively large eyes; unfortunately, the crushing of this area makes it difficult to determine if the eye sockets were elevated above the rest of the skull, as in *P. uncia*, to give it the effective binocular vision and depth perception that are advantageous for mountain hunting. Its features indicate that *P. blytheae* is likely a direct ancestor of the living *P. uncia*, previously considered the most basal living member of the genus *Panthera*. In life and in its habits it may have resembled the living Andean mountain cat, *Leopardus jacobita* (plate 3).

The now endangered snow leopard *Panthera uncia* (figure 3.10) is, like the clouded leopard *Neofelis*, only distantly related to the common leopard

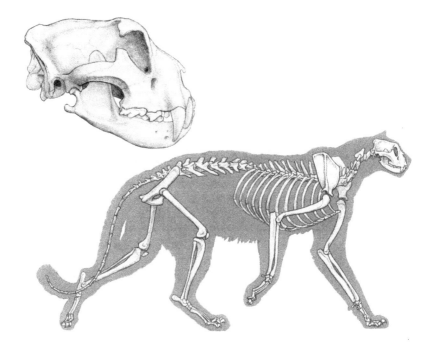

FIGURE 3.10. Skull and skeleton of snow leopard, *Panthera uncia*.
In addition to body proportions that make it such an excellent climber and leaper, the snow leopard's skull has higher orbits (eye sockets) and more widely spaced eyes than the skulls of other pantherins, which gives it exceptional binocular vision for judging distances during the vertical pursuit of bharal, ibex, and other prey. The short, powerful upper and lower jaws provide great power in making quick, killing bites.

Panthera pardus. The snow leopard is a distinct lineage that, along with tigers, diverged from other big cats between the late Miocene and early Pliocene (see chapter 4). Anatomical features and molecular analyses of living *Panthera* species place the clouded leopard *Neofelis* very close to the basal Pantherini (or Pantherinae), which may have separated from the rest of Felidae as early as 16.4 Ma. This split, also indicated by molecular studies of living forms, occurred long before the radiation within *Panthera* that gave rise to the more closely related leopard-jaguar-lion group.

Snow leopards, whose males weigh between 25–55 kg (55–121 lb.), are still widespread but now sparse and rarely seen in the mountain ranges of central and southern Asia, where they generally inhabit alpine and subalpine zones from elevations of 3,000 m (9,800 ft.) to over 5,000 m (16,404 ft.). These cats are amazingly agile hunters on rocky, precipitous slopes (figure 3.11), and their prey includes a variety of small (hares) to large (ibex) animals, the most important of which, on the Tibetan Plateau and to the north, is the agile bharal or "blue sheep" (*Pseudois nayaur*), a basal caprin displaying body features and behaviors that make it close to the common ancestry of both goats and sheep. This caprin is goatlike in its general preference for high cliff habitats—unlike sheep, of which most living species tend to favor less steep terrain. The bharal is a sustaining, vital species for the snow leopard, and the naturalist and wildlife biologist George B. Schaller (Wildlife Conservation Society) and others believe that its ancestors, as well as those of today's other endemic Tibetan rupicaprines, have a long prey-predator relationship going back to the time of *P. blytheae*. Bharal are substantial: the males may weigh about 60–75 kg (132–165 lb.); in the western ranges, the snow leopard's other common large prey, the Himalayan ibex (*Capra sibirica*), can weigh 80–100 kg (176–220 lb.). Thus it seems likely that *P. blytheae* or a close relative began to evolve a larger body size in order to tackle the Tibetan Plateau's ruminants as they, too, increased in stature and weight. *P. blytheae* and its descendants became predators of agile, mountain-dwelling, progressively larger-bodied rupicaprins and caprins. From the Pliocene onward, the pantherins' prey base on the Tibetan Plateau expanded to include the plains-living ovibovins, sheep, saigin antelope, and, later, other ruminants.

The origin of genus *Panthera* is, however, only part of a much bigger picture. In a real sense the Tibetan Plateau, whose altitude fosters low temperatures that rival that those of the high-latitude Arctic and Antarctic, can metaphorically be considered the earth's "third pole." Accordingly, it has produced animals that are highly cold adapted and, in the case of carnivorans, have become even more predatory and survive by increasing their caloric intake. A series of recent fossil discoveries has led Tao Deng (Institute of Vertebrate Paleontology and Paleoanthropology, Chinese Academy of Sciences), Wang, and their associates to suggest that the

FIGURE 3.11. Snow leopard pursuing bharal.
In this sequence, a snow leopard displays its remarkable agility in pursuing
and capturing prey, a young female bharal or "blue sheep," down a steep cliff.
The flexibility of the spine in ancestral pantherins was crucial in enabling them to
hunt the equally agile rupicaprins and caprins that were adapting to the elevation
of the Tibetan Plateau and was key to their later success as ambush predators of
large ungulates.

Tibetan Plateau region, with its endemic faunas that could withstand extremely frigid, dry conditions, was in a sense a "training ground" for the evolution of later, far more cold-adapted types that would radiate from the Tibetan Plateau to become some of the most iconic, typical megaherbivores of the Pleistocene **glacial** periods (Deng 2011).

Among the best known Pleistocene species are the woolly mammoth (*Mammuthus primigenius*) and woolly rhinoceros (*Coelodonta antiquitatis*). These two animals, with their long, insulating outer coats, fleecy underwool, fat deposits, and cranial appendages (tusks and horns) for sweeping aside snow to uncover forage, were particularly well suited to survive long, frigid winters, and it was previously thought that such features evolved in response to the Eurasian ice sheet expansions during the early Pleistocene. In 2007, however, the complete skull and associated postcranial bones of a new woolly rhino species (*Coelodonta thibetana*) were discovered by Wang and his team in the Zanda Basin in a middle Pliocene stratum dating to ~3.7 Ma, well before the beginning of the first Pleistocene glacial, about 2.8 Ma. Although no soft tissues were preserved to prove it had evolved insulation from cold, the new rhino possessed more basal skull features and lower crowned teeth than the subsequent, and more derived, early and middle Pleistocene species, *Coelodonta nihowanensis* and *Coelodonta tologoijensis*. The adaptations that permitted *C. thibetana* to exploit the Tibetan Plateau's high altitude enabled its descendants to disperse into higher latitudes as the climatic cold intervals intensified (figure 3.12). The discoveries of rhino species in subsequent geological epochs as well as in higher latitudes strongly suggest that *C. tologoijensis* was probably the direct ancestor of the late Pleistocene *C. antiquitatis*. This was the well-known, late Pleistocene woolly rhino species that ranged over the entire Eurasian landmass from the British Isles and southern Spain to northeast Asia and was portrayed on cave walls by our human ancestors.

The discovery of this ancestral woolly rhino, in combination with other new species, gives great support to Deng and associates' hypothesis that several important cold-tolerant ice-age mammal lineages evolved from earlier endemic Tibetan regional forms, a concept they informally term the "out of Tibet" hypothesis. In addition, other new fossil discoveries suggest that that besides being a "training ground" for later ice age herbiviores, the cold environment imposed metabolic demands that forced carnivorans, too, to adapt by becoming even more specialized in eating flesh. Three fossil canids, the ancestral "hunting dogs" *Xenocyon* sp. and *Sinicuon* (cf. *S. dubius*), plus the ancestral arctic fox *Vulpes* (*Alopex*) *qiuzhudingi*, all show modifications in their teeth that indicate the transition from a mesocarnivore diet to that of a hypercarnivore. In the hunting dogs, certain cusp structures like the trigonid of the first lower molar became tall, narrow, and sharply bladed, whereas in the fox a slightly different variation

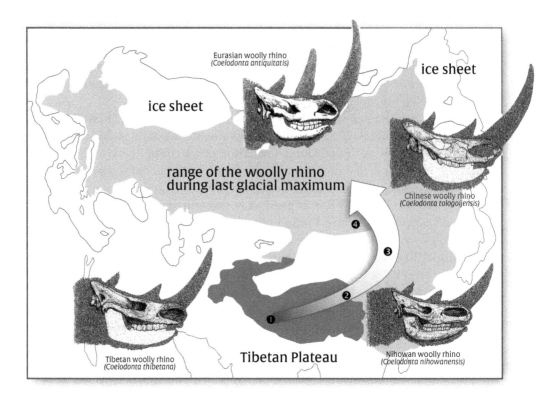

FIGURE 3.12. Evolution of the woolly rhinoceros.
Dispersing from their ancestral point of origin on the Tibetan Plateau (arrow),
woolly rhinos were preadapted to life in northern latitudes as these became
progressively cooler. The Early/Middle Pliocene Tibetan woolly rhino *Coelodonta
thibetana* (locality 1) is currently the earliest representative of the lineage,
which evolved into the Early Pleistocene Nihowan rhino *Coelodonta nihowanensis*
(localities 2, 3) and later into the Chinese rhino *Coelodonta tologoijensis* (locality 4),
finally culminating during the Late Pleistocene with the iconic Eurasian woolly
rhino, *Coelodonta antiquitatis*, which had a range spanning from Spain to China.
Source: After Deng et al. (2011).

occurred. Both modifications served to cut flesh more efficiently and
converged on felid adaptations (figure 3.13). The changes in these animals' teeth reflected increased caloric needs for surviving in frigid winters,
needs that could be best satisfied by consuming more meat.

Today, terrestrial carnivorans that inhabit high arctic regions like the
arctic fox (*Vulpes* [*Alopex*] *lagopus*), Eurasian wolf (*Canis lupus*), and polar
bear (*Ursus maritimus*) are all almost exclusively carnivores but evolved from
mesocarnivores whose diets included some plant matter. The newfound
Tibetan fossil hunting dogs predate the existence of modern, low-latitude

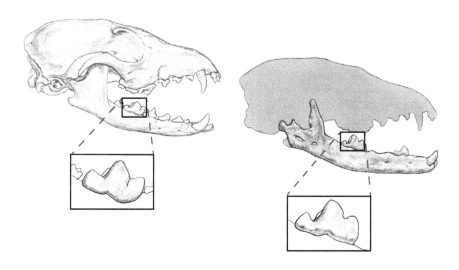

FIGURE 3.13. Changes in arctic fox dentition.
Compared to the first lower molar of the living arctic fox (*Vulpes* [*Alopex*] *vulpes*)
(*left*), that of the fossil form (*Vulpes* [*Alopex*] *qiuzhudingi*) (*right*) displays a taller,
more pointed protoconid or middle cusp. This was an adaptation for slicing flesh
more efficiently, which parallels the similar condition in felids and indicates a
transition to a more flesh-based—and more calorie-rich—diet for arctic foxes in
colder northern latitudes.

hunting dogs, the Asian dhole (*Cuon alpinus*) of south Asia and the Cape
hunting dog (*Lycaon pictus*) of Africa, which in turn says something else:
that the hypercarnivorous lifestyle of the Tibetan canids was subsequently
adopted by canids at warmer, lower elevations and latitude. This had tre-
mendous consequences for felids, too.

As the endemic caprin and rupicaprin bovids became increasingly
larger-bodied, the early pantherins underwent a correspondingly comparable increase in size (figure 3.14). With the advantages these cats already
held in overcoming smaller prey, larger size would maximize the amount
of meat they could obtain for the effort they expended. Although big cats
often have preferred prey species, they are nevertheless highly opportunistic.
If fossil pantherins were mainly solitary predators, as many of today's
forms are, their need for huge hunting territories to sustain their needs
would mean that they ranged as widely as today's forms, bringing them
into contact with large potential prey outside the Tibetan Plateau. In the
less steep, hilly plateau regions of the transitional Miocene-Pliocene this
would have originally included antilopin bovids like the ancestral chiru
or Tibetan antelope (*Qurliqnoria cheni*) and *Antilospira licenti*, early sheep

FIGURE 3.14. Size relationships of pantherins and their prey.
Selected Late Pliocene to Late Pleistocene ruminants and the range in size of
pantherins that were likely to have preyed on them.

(*Protovis himalayensis*), and the ancestral Eurasian wild pig (*Sus* sp.) and horse (*Hipparion [Plesiohipparion] huangiense*), among others. Tackling prey with larger body size, however, would bring them into competition with the machairodontine sabertooths that for many millions of years held the monopoly as hunters of larger ruminants and other big game. What would happen when pantherin met sabertooth?

CHAPTER 4
Beyond the Distant Horizons

Qaidam Basin, northeast China, 3.7 million years ago: Grunting loudly, flailing its horns and stamping to show its continuing strength and defiance, the bloodied *Leptobos brevicornis* challenges the would-be predator. Although smaller than a modern bison the male is in his prime, and he easily shook off awkward attempts by the young *Panthera palaeosinensis* to make a kill by mounting its prey's shoulders and trying to bite into the thick nape of the neck. As the nighttime standoff continues, a second cat and then a third emerges from the thicket along the snow covered riverbank: two female dirk-toothed sabertooths, *Megantereon cultridens*, their protruding upper canines shining in the moonlight. Within two seconds they sprint forward and grapple the bison. Gripping the bison's shoulders the sabertooths wrestle it down with their powerful forelimbs and shortened backs, dispatching it with an expertly delivered shear-bite. Turning only long enough to warn the inexperienced pantherin away with snarls, both begin to feed (figure 4.1).

. . . . .

As the now larger early pantherins began to disperse from the Tibetan Plateau in search of prey during the course of the Pliocene, scenes such as the this were probably common. For the relatively small, early felines the test, bite, and pierce technique for killing prey at the nape of their necks worked well. However, using this technique on larger ungulates such as deer, bison, and horse was often unsuccessful because of the massiveness of their necks. Prey animals of this size and larger were the province of the leopard- to lion-sized saber-toothed machairodontines. To compete with those master predators the early pantherins had to develop new techniques for killing besides simply becoming larger. Like the sabertooths, they were capable of capturing and successfully wrestling a prey with the forelimbs, but they lacked the long upper canines that made the shear-bite so successful. Instead, the pantherin's shorter canines were used to either to clamp around the upper throat or to cover the nose and mouth, both of which shut off air and produced asphyxiation (figure 4.2). This took much longer than a shear-bite to kill the prey, as much as ten minutes, but with less risk of breaking the canines.

FIGURE 4.1. *Megantereon* sabertooths warning an early pantherin to stay away from a kill.

FIGURE 4.2. Tracheal and nose clamping methods to kill prey.
To kill large ungulates such as this topi (*Damaliscus lunatus jimela*), lions and other pantherins typically revert to either clamping their jaws over the nose (*above*) or across the throat (*below*) to asphyxiate their prey, usually in about five minutes.

Field observations of tigers and lions by Schaller and others show that while a nape bite is often still the choice for smaller deer and antelope, muzzle or neck clamping is commonly used to kill big ungulates. Others have reported seeing these cats wrestle a domestic river buffalo (*Bubalus bubalis bubalis*)'s head close to the ground in the opposite direction to its line of fall, which sometimes results in dislocating or even breaking the neck.

Trial and error can and does play a part in teaching an individual big cat which predation technique works and which does not. The killing skills taught by the mother before the offspring matures were also vital (figure 4.3). In the case of modern tiger mothers and cubs, the period of training starts when the cubs are about five or six months old. It can sometimes take up to sixteen months for the cub to become a self-sustaining adult potentially capable of killing its own large prey. At this point the cubs are nearly as large as their mother and often begin to hunt on their own but are sometimes clumsy when attacking larger animals. This ineptness can result in serious injury to the cub or worse. Even experienced adult tigers face great

FIGURE 4.3. Mother tiger teaching cubs to kill.
When the cubs are old enough (about five or six months), the mother begins
teaching them how to hunt and kill by bringing to the den small animals that
she has disabled (shown here, the giant squirrel *Ratufa indica*) so that the cubs
can practice pouncing and grappling with them. Because they are too young and
inexperienced to make the actual kill, the mother must demonstrate this by
a final bite to the prey's neck.

danger when they choose as their prey brown bears or Eurasian wild pigs,
which are fast and powerful and can use their claws or razor-sharp tusks
to stun, disembowel, and kill their attacker. Almost-grown tiger cubs may
remain longer with their mother (for females, an additional four to six
months) to solicit food before seeking their own territories as adults.

Although larger body size helped early pantherins tackle bigger
bovids, it also posed a problem. While a large warm-blooded predator
requires proportionately fewer calories than a small one on a daily basis
to sustain its daily metabolic needs, a minimum amount of food is still
required. Although modern large pantherins such as tigers and lions are,
like other carnivorans, opportunistic and will prey on any animal they
can catch, there is a tendency to try for larger game because more meat
means an increased caloric payoff in exchange for the effort: a rabbit is
just a snack but a buffalo can last for days. However, any region's prey
base, even if composed of several available larger species, can only sustain a given number of hypercarnivorous predators over the long term.
This is why most large felines tend to be solitary as adults (although not

always, as with lions, occasionally tigers and cheetahs, and, in the past, probably some sabertooths), and they survive by having extensive, often enormous hunting territories. One snow leopard adult may require, on average, a territory roughly the size of New York City's greater metropolitan area (13,318 sq. mi.) each week to find enough prey, large and small, to sustain itself; each day it must cross and recross its range to find food. As a result, populations of early pantherins gradually dispersed across the regions surrounding the Tibetan Plateau in order to encounter new and diverse prey species. They also competed for the now increasingly bigger bovids, cervids (deer), and other ungulates exploiting the mosaic of open woodlands and savannas with the felid sabertooths and the continent's other specialized carnivorans, including perhaps the last of the Late Miocene giant hyenas (*Dinocrocuta gigantea*) and the "dog-bear" hemicyonids. As the Pliocene progressed the bovids became increasingly diverse and, along with some of the perissodactyl lineages such as horses, provided new prey resources for both types of cat and for other carnivorans.

At some time during the mid- to late Pliocene the first major diversification or split within the genus *Panthera* is believed to have occurred. Molecular studies, based on living pantherin DNA from blood and skin samples, indicate that snow leopards and tigers form a separate clade that became distinct from a more generalized jaguar-leopard-lion lineage as early as 3.60 Ma. The reasons for this divergence are not clear at this time but could relate to an earlier dispersal of the snow leopard–tiger clade into northern China while the then more generalized jaguar-leopard-lion group remained longer in the parent region of the Tibetan Plateau.

Apart from the most obvious but superficial differences of coat pattern and size, tigers, lions, leopards, and jaguars are very similar anatomically, distinguished by relatively minor variations in their skeletons with respect to the extent and position of craniofacial sutures, skull proportions (figure 4.4), and metacarpal bones. These, however, suffice to separate the snow leopard–tiger group from the jaguar-leopard-lion group. Although they have no formal taxonomic recognition, these informal clades provide a discernible pattern for evaluating the phylogeny of fossil pantherins. As of this writing we have no fossils with combined snow leopard/tiger characters to corroborate the molecular findings, but both living species share similar craniofacial skull suture patterns and limb proportions. If the hypothesis of the dispersal of larger, cold-adapted pantherins into the northern latitudes of Asia is correct, however, this makes northern China a likely area for the discovery of later cats, and we should expect a fossil cat combining snow leopard and tiger features eventually to be found there. As for snow leopards, apart from the skull of the probably ancestral Tibetan snow leopard *Panthera blytheae* dated to about 4.4 Ma (but whose geological range could extend as far back as 5.95 Ma), few pre-Holocene bones of this cat are known. A snow leopard relative (*Panthera* cf. *P. uncia*)

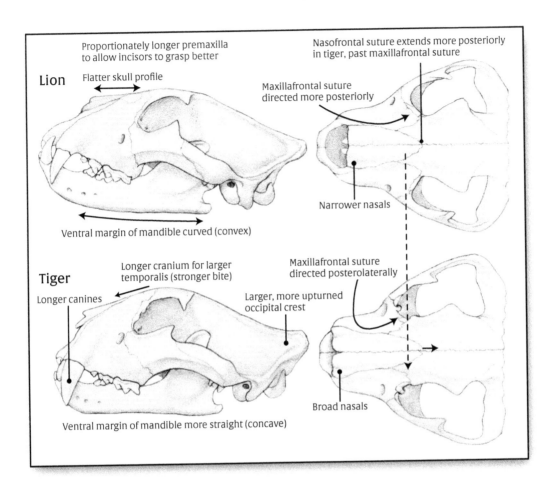

Lion

Proportionately longer premaxilla to allow incisors to grasp better

Flatter skull profile

Nasofrontal suture extends more posteriorly in tiger, past maxillafrontal suture

Maxillafrontal suture directed more posteriorly

Narrower nasals

Ventral margin of mandible curved (convex)

Tiger

Longer cranium for larger temporalis (stronger bite)

Longer canines

Maxillafrontal suture directed posterolaterally

Larger, more upturned occipital crest

Broad nasals

Ventral margin of mandible more straight (concave)

FIGURE 4.4. Pantherin craniofacial skull sutures.
Although the skulls of big cats like a lion (*above*) and a tiger (*below*) look much alike, subtle differences exist that reflect their different styles of predation. The lion, a group hunter, has a slightly extended muzzle and a distinctive cranial-suture reinforcement to accommodate the stresses in biting and holding large prey while other pride members join in to restrain and make the kill as a group. The tiger, a solitary hunter, depends on a shorter muzzle, larger canines, and greater anchorage for jaw-closing muscles to produce a stronger bite in killing prey more quickly, requiring a different suture-reinforcement pattern to make the skull stronger.

has been reported from locality 73 in the Pabbi Hills, Pakistan, with a mid-Pleistocene date of 1.2–1.4 Ma, but its snow leopard identity may be problematic.

Although genetic evidence suggests earlier divergence from a common pantherin ancestor, the earliest known tiger appears in the early Pleistocene of northern China. ***Panthera zdanskyi*** (figure 4.5) is based on an

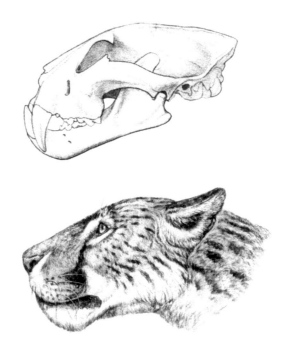

FIGURE 4.5. Skull and life restoration of *Panthera zdanskyi*.
Known from a well-preserved cranium and mandibular rami from in China, this
cat, informally named the "Longdan tiger," has a variety of skull and tooth features
that are consistent with its being the earliest tiger species.

extremely complete, slightly laterally compressed cranium and associated
mandibular rami found in the Longdan loess deposits of Gansu Province,
northwest China, and dates to the Gelasian stage of the early Pleistocene,
about 2.55–2.16 Ma. This **holotype** specimen and the paratype premaxillae
and partial maxillae (upper jawbones) formerly attributed to *P. paleosinensis*
(see later in this chapter) display skull characters found in fossil and modern tigers—a heart-shaped narial aperture, long nasals relative to the skull
size, a nearly straight lower border of the mandibular ramus (lower jaw),
and numerous dental characters—even though the first specimens of
Panthera tigris do not occur in the fossil record until half a million years
later. The skull bears a mosaic of features suggesting that various parts of
the cranium, mandible, and dentition were changing at different rates, a
condition also seen in some fossil cheetahs and commonly seen in other
rapidly evolving forms. In the case of *P. zdanskyi*, the robust skull, large
canine teeth, and size of the upper postcanine teeth are decidedly tiger-
like, whereas some aspects of the mandible and lower postcanine teeth are
relatively more basal than that of later true tigers but "caught up" with the
cranium later in tiger evolution. Informally named the "Longdan tiger,"

this cat's characters more closely resemble those seen in some Southeast Asian tiger subspecies than those of mainland Asia subspecies, which could mean that by this time a further diversification had already occurred in the most ancestral tigers.

The oldest specimen attributed to the extant species *Panthera tigris*, from Sichuan, China, is a little younger (~2.0 Ma) than *P. zdanskyi*. *P. tigris acutidens* (figure 4.6, plate 4a, b) is known from two well-preserved adult skulls and several fragmentary juvenile skulls discovered in Sichuan, China, along with adult limb and metapodial bones. Named the "Wanhsien tiger" after its discovery locality in Yen Ching Kao, Wanhsien, the bones are anatomically similar enough to assign it to the extant species, *P. tigris*, but sufficiently different for it to be considered a fossil subspecies. As large as or even larger than the extant Amur (Siberian) subspecies *P. t. altaica*, the Wanhsien tiger weighed about 300 kg (660 lb.), probably had a widespread northern range, and is also known from the 230 ka Zhoukoudian (Chouk-outien) locality in China. Other maxillary and mandibular fragments are

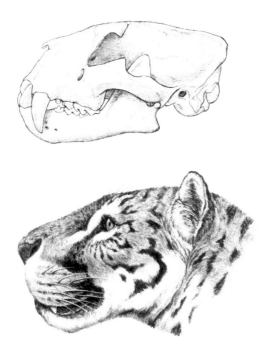

FIGURE 4.6. Skull and life restoration of *Panthera tigris acutidens*.
Known informally as the "Wahnsien tiger," this form predates *Panthera zdanskyi* but is the earliest tiger to represent the modern species *Panthera tigris*. It had a robust skeleton that rivaled the extinct "Ngandong tiger" from Java and the recent Amur (Siberian) tiger in size (see also plate 4a, b).

known from the Calabrian stage (~1.8 Ma–781ka) of the early Pleistocene of Lantian, China, and are believed to represent a tiger but are not sufficiently diagnostic to be assigned to a species.

The trail of fossil tigers now changes abruptly from northern China to the 1.3 Ma–700 ka, Middle Pleistocene Trinil locality in Java, Indonesia, the same riverbank site made famous by the discovery of "Java Man" (originally *Pithecanthropus erectus*, now *Homo erectus soloensis*) by Eugéne Dubois. *Panthera tigris trinilensis*, the "Trinil tiger," is known mainly from a left mandibular ramus, which, although massively robust, probably represented a medium-sized tiger of about 110–150 kg (242–330 lb.). The sudden shift to Indonesia is probably in part caused by a lack of discoveries in China that would geographically connect presumed subspecies with *P. tigris trinilensis*. Another recognized, large fossil subspecies based on seven incomplete specimens, *Panthera tigris soloensis*, was discovered in the same region and is named the "Ngandong tiger" for its proximity to Ngandong, Indonesia. During glacial intervals of the Pleistocene, Java and the other islands of modern Indonesia were united by lowered sea levels into a huge peninsular subcontinent, larger than India, called Sundaland. Occasionally punctuated by volcanic eruptions, subtropical to tropical conditions prevailed over a mosaic of savannas, rainforests, and mesic woodland where large fossil ungulates like the water buffalo (*Bubalus palaeokerabau*), banteng (*Bos palaeosondaicus*), and occasionally juvenile proboscideans (*Stegodon trigonocephalus, Elephas hysudrindicus*) would have provided a partial prey base for the endemic tigers, along with the extant Malayan or Asian tapir (*Tapirus indicus*) and Sumatran rhino (*Rhinoceros sondaicus*). According to genetic evidence, neither *P. t. trinilensis* nor *P. t. soloensis* was the direct ancestor of the living Sumatran and now extinct Balinese tigers in spite of their location (see chapter 8).

The story of tiger evolution has some surprises. Molecular analyses in 2016 by David M. Cooper (University of Edinburgh) and his associates showed that at the start of the Holocene **interglacial** the tiger's core distribution was in southern China and eastern Sundaland rather than central Asia and, rather unexpectedly, that these cats actually dispersed into Central Asia from Southeast Asia instead of the other way around. This explains why the critically endangered—and possibly now extinct in the wild—South China tiger (*P. t. amoyensis*) is genetically and anatomically the most basal of all modern tigers and the ancestor of all the other subspecies. During early interglacial periods low sea levels decreased the distances between now more separated islands like Java and Bali, and tigers, notably good swimmers, were able to cross the narrow channels to recolonize these areas. But why *recolonization*, when these areas should already have had suitable tiger habitats and populations for thousands of years? The answer lies in the catastrophic prehistoric eruption of huge Mt. Toba, Sumatra, about 73 thousand years ago, the effects of which were widespread. The largest volcanic event in the past 2 million years, the Mt. Toba super-eruption dispersed huge volumes of ash across much of the Indian Ocean, Indian peninsula, and South China Sea and may have caused a "bottleneck" or genetic constriction in the dispersal of early humans out of Africa. The eruption apparently eradicated local Indonesian tiger populations; the subsequent global cooling and drought in the decades following the eruption negatively affected tiger habitats throughout most of Sumatra and the Malay Peninsula (figure 4.7). As a result these lands temporarily lost their tigers and only later were repopulated. During the Pleistocene both Asian

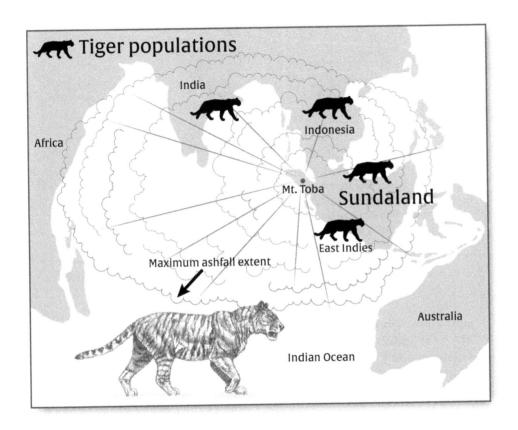

FIGURE 4.7. Geography of Sundaland, showing tiger populations and Mt. Toba eruption.
Lowered sea levels during the Weichselian Glacial facilitated the emergence of Sundaland—an extensive area of land that incorporated present day Indonesia and the East Indies. The eruption of Mt. Toba in northwest Sundaland (now Sumatra) at about 73 ka had a catastrophic effect on surrounding ecosystems, resulting in the local extinctions of some species, including tigers. Although tigers later gradually recolonized these areas, the event produced a severely decreased genetic variability or "bottleneck" in tiger and other animal populations.
Source: After Luo et al. (2004).

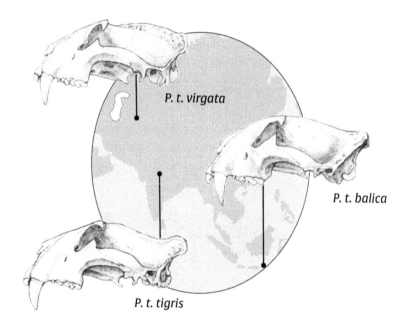

P. t. virgata

P. t. balica

P. t. tigris

FIGURE 4.8. Tiger subspecies comparison based on skull morphology.
During the Pleistocene tigers were widely dispersed throughout Asia. Although
recent genetic studies now suggest the existence of only two subspecies, a
"Continental" (Asian mainland) and a "Sunda" (largely island) form, previous
anatomical studies recognized a number of subspecies based on skull morphology
and other characters. Some of these, such as *Panthera tigris tigris, P.t. virgata* and
P. t. balica, display considerable variation in the occipital crest, implying differences
in muscular and skeletal adaptations. *Source:* Skulls modified from Mazák (2010).

and Indonesian mainland tigers adapted to a variety of biomes, resulting
in several distinctive skull morphologies, which have been documented by
Kitchener and J. H. Mazák (figure 4.8). In the past cranial characters and
geographic locations have been used as the basis for assigning subspecific
names for tigers, but some workers, in light of recent genetic studies, now
prefer to recognize only two subspecies (see chapter 7).

Although the paleontological record for the origins of the earliest
and most definitive jaguar-leopard-lion-like pantherins is currently poor,
the results of molecular analyses suggest that the divergence of jaguars
from the monophyletic group of lions-leopards may have occurred at ~2.56–
3.66 Ma and lions and leopards diverged from each other at ~1.95–3.10 Ma.
The earliest paleontological evidence for a basal jaguar-leopard-lion-like
pantherin, **Panthera palaeosinensis**, is based on a well-preserved holotype
skull and left hemimandible found in Henan Province, northern China
(figure 4.9). Although missing the upper canines, the robust skull and
left lower jaw are otherwise complete and show a mosaic of characters

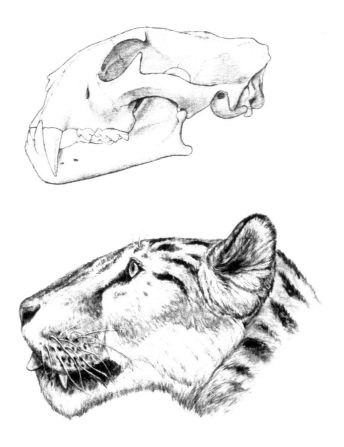

FIGURE 4.9. Skull and life restoration of *Panthera palaeosinensis*.
This species is currently recognized as the earliest known representative of the
pantherin lineage that includes jaguars, leopards, and lions after this group
split from the snow leopard–tiger lineage. Some workers interpret it as a very
basal leopard.

that originally produced much discussion as to their phylogenetic status.
Some workers proposed that *P. palaeosinensis* was a either a small basal
tiger or a species close to the ancestral tiger, but Mazák (Shanghai Science
and Technology Museum), Per Christiansen (Aalborg University), and
Kitchener demonstrated in 2011 that both attributions were incorrect.
Others pointed out that the mosaic combination of various pantherin
features seem to fit well with its being at least a basal form. In a 2010
study that involved a multivariate analysis of 504 fossil and living
pantherin skulls (in which adjustments were made for size), Mazák and
other colleagues found certain features that suggested *P. paleosinensis*
could be a representative of either a very basal leopard or an ancestor close to the jaguar-leopard-lion lineage following the divergence

of the snow leopard–tiger group. A critical problem in evaluating the place *P. palaeosinesis* should occupy in early pantherin phylogeny is its date—reported in some scientific literature as being very late Pliocene (European Villanyian / Asian Mazeguoan LMS, both c. 3.60 Ma) or earliest Pleistocene in age and by others as possibly *mid*-Pliocene or older (Mazák et al. 2010). If correct, the latter would fit better with the accurately dated, later age of the Laetoli cats (see later in this chapter) and would, along with its basal pantherin characters, place *P. palaeosinesis* in time before the next oldest known pantherin, **Panthera gombazoegensis** (or, according to some workers, *P. onca gombazoegensis*).

Panthera gombazoegensis is known from a variety of incomplete skull specimens and postcranial bones from the Early Pleistocene of Eurasia (~1.95–1.77 Ma) (figure 4.10). There is broad agreement as to its overwhelmingly jaguar-like affinities, which is why this cat, in spite of its age, is

FIGURE 4.10. Skull and life restoration of *Panthera gombaszoegensis*.
Eurasian jaguars occurred in cave localities throughout Europe and Asia from the Villafranchian through the Holsteinian interglacial stages of the early Pleistocene. The fossil form's great similarity to modern jaguars suggests that it hunted prey in open woodlands and riverine gallery forests.

sometimes referred to as a fossil subspecies of the modern jaguar. While Alan Turner recognized all the European specimens as belonging to a single species, *P. gombazoegensis*, that was probably ancestral to the living jaguar and its current subspecies, others, such as Helmut Hemmer and Hannah O'Regan (University of Nottingham), refer *P. gombazoegensis* to the extant *P. onca.* In 2010, after a multivariate analysis, Hemmer and colleagues proposed that a newly recognized fossil west Asian jaguar from the Caucasus Mountains region, which he described as *Panthera onca georgica* from Dmanisi, Republic of Georgia, was probably an intermediate form between the earlier, European *P. o. toscana* and two later subspecies, the late Pliocene–early Pleistocene *P. o. gombaszoegensis* and the later Pleistocene *P. o. augusta* of North America. This conclusion is based on an evaluation of *P. o. georgica*'s carnassials that appear to be transitional between the sectorial features of *P. o. toscana* and the carnassials of *P. o. gombaszoegensis* and that show both a cutting and a crushing morphology. This is a highly distinctive and unique dental reversal in a felid but fits with the current predatory behavior of some modern South American jaguars, whose extremely large heads and powerful jaws actually give adults the ability to prey on river turtles by crushing their shells (figure 4.11). For the Eurasian subspecies, tooth characters in association with fossil localities suggest that a stem population of jaguars dispersed from a probable Asian point of origin to both Europe and North America, producing at different times the subspecies *P. o. augusta* and probably other, South American subspecies, which survived from 1.8 Ma until 11ka. The large, extinct North American *P. o. augusta* does not occur in the fossil record until the Rancholabrean stage of the Pleistocene (195 ka) and is considered to be ancestral to all extant South American jaguars.

During the mid- to late Pliocene East Africa may have shared the spotlight for pantherin origins with northern Asia. In Tanzania maxillary and mandibular remains, a few isolated tooth fragments, and some postcranial elements of presumed pantherins were found in deposits of the Upper Laetolil Formation, Laetoli, where radiometric dating produces a (Middle Pliocene) age of 3.8–3.5 Ma. The incomplete nature of these specimens makes it impossible to determine anything other than that they represent lion- and leopard-sized individuals and, as we learned in chapter 1, estimated body size based on biometrics alone is not always a reliable indicator of species status. In keeping with the historic controversies over extinct lion phylogeny (see later in this chapter) there is disagreement over the status of the Laetoli specimens. Although one worker, Hemmer, identified them as early pumas, Werdelin and Lewis (2013) interpret the Laetoli fossils as showing significant differences from extant lions and leopards but note that thereafter lion-sized and leopard-sized pantherins are fundamental elements of the carnivoran communities of Africa and Eurasia. Lion- and leopard-sized pantherins have been recovered from localities older

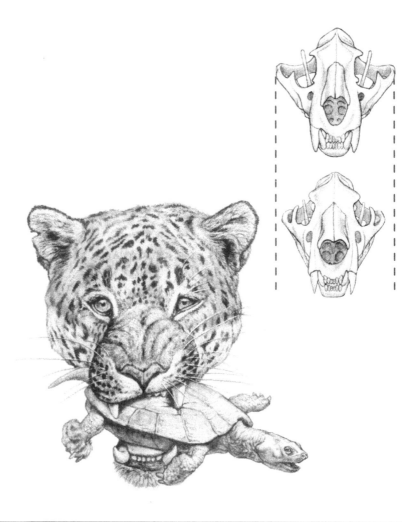

FIGURE 4.11. Modern jaguar crushing the shell of a young giant South American river turtle, *Podocnemis expansa*.
The skull of a jaguar (*upper right*), drawn to the same absolute width as that of a tiger (*lower right*) has wider zygomatic arches that provide great mechanical advantage for the masseteric muscles and allow jaguars to routinely prey on river turtles and caimans. Although a jaguar's bite is only 75 percent as strong as an adult tiger's in terms of absolute pounds per square inch, for its size and weight a jaguar's bite (750 psi) is relatively stronger because of this muscular advantage and its shorter muzzle.

than 3 Ma in the Lake Turkana Basin of Kenya and Ethiopia. The earliest occurrence of the extant leopard ***Panthera pardus*** is about 2.8 Ma from the South African locality of Sterkfontein, whereas the extant lion *Panthera leo* is widespread at localities dated at 2 Ma or younger. Lions and leopards became increasingly common in Africa after about 1.8 Ma, coinciding with the demise of their sabertoothed relatives. The location and early age of these finds demonstrate that big cats had already dispersed far beyond Asia by the late Pliocene, and it is highly possible that future discoveries may lie along the migration routes between the two continents.

Recently the skull of a giant lion was discovered in Member I of the Kibish Formation in the Natodomeri area at the Kenya–South Sudan border. It has a Late Pleistocene date of between 195 and 205 ka, and Fredrick K. Manthi (National Museums of Kenya) and his associates have referred it to *P. leo*, in spite of an inferred size that was far greater than any other known individuals of this species and equal to the largest **Panthera spelaea** (see later in this chapter). These authors offer at least three hypotheses to explain the presence of such a huge individual modern lion at this **geochron:** first, it represents a species distinct from although closely related to *P. leo*; second, its unusually large body size correlates with the cooler climate that occurred at that time in that region; and finally, it belongs to an extinct (and distinctive) population or subspecies of *P. leo* larger than any modern representative and, as seen in chapter 3, its size resulted from competition with other large predators. Yet another possibility is a twist on the first hypothesis: the lion's huge size may indicate that it was indeed a different species from *P. leo* and that it represents a dispersal of some populations of the Eurasian steppe lion *Panthera spelaea* into northeastern Africa during glacial episodes when lowered sea levels would have produced a connecting land corridor.

The oldest reported leopard fossils are from late Pliocene and early Pleistocene sites in India and Africa. Interestingly, phylogenetic trees based on molecular DNA studies indicate an African, not Asian origin for all the extant leopard *Panthera pardus* subspecies younger than middle Pleistocene, although that does not rule out an Asian origin for the species. As many as fifteen extant African subspecies were formerly recognized on coat patterning and cranial differences, but recent genetic evidence indicates that all African leopards belong to a single subspecies, *Panthera pardus pardus*. The African subspecies is considered the most basal, with those from western and central Asia the next most recently evolved and those of northern and eastern Asia the youngest of all; this pattern is an intriguing reversal of the "out of Asia" origin that seems to be the general trend in pantherin dispersals. Although usually associated with Africa and Asia, leopards occurred across Europe during the Pleistocene and well into recent or Holocene time, with late finds in Spain, Greece, and Ukraine dated to the first century CE. As many as four Pleistocene subspecies are recognized by some workers, with the Late Pleistocene *P. pardus spelaea* the largest. Its cranial sutures most closely resemble those of the extant Persian leopard *P. pardus ciscaucasica* (= *P. p. tulliana, saxicolor, dathei*), but the skull is morphologically different from those of some extant leopards (figure 4.12). Where male and female skulls are known, these display

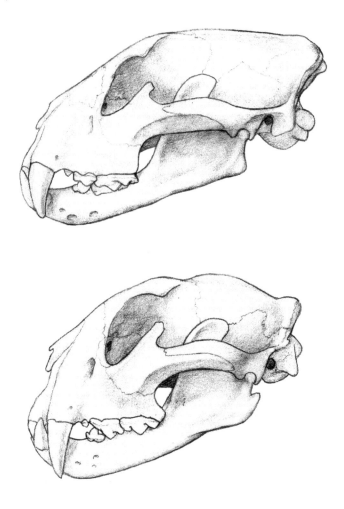

FIGURE 4.12. Skulls of modern African leopard *Panthera pardus pardus* (*above*) and Pleistocene European leopard *Panthera pardus spelaea* (*below*).
Known from cave deposits across Europe and western Asia, Eurasian leopards were larger than their modern counterparts and had shorter skulls.

pronounced **sexual dimorphism**. Other aspects of the "cave leopard" are currently unknown, but the "Red Leopard" drawing (plate 6a) in the Chauvet-Pont-d'Arc Cave of the Ardèche region of France offers a tantalizing glimpse of a spotted coat that contrasts with a pure white belly, differing from that of any extant leopard.

Pleistocene Eurasian leopard fossils are generally rarer than those of steppe lions, but their habitat preferences and predation habits can be inferred. Ice-age leopards had a wide distribution and probably preferred mountain/alpine boreal forest to **mammoth steppe**. In addition to using caves as safe refuges to bring their kills for uninterrupted consumption, leopards may, like steppe lions, have attempted winter predation on the hibernating cubs of brown or cave bears when other prey was scarce (see chapter 5).

At the very late (12,000 ka) Pleistocene cave locality of Los Rincones, Spain, the remains of several cave leopards have been recovered, making up 12.2 percent of the known carnivoran mammal fossils there, second only to the most common carnivoran from this locality, the brown bear (*Ursus arctos*). Nearly two-thirds of the locality's bones are of the Spanish ibex (*Capra pyrenaica*), and their relatively intact state suggests that leopards rather than hyenas were the main predators of this ungulate and that the leopards brought their prey back to the cave for consumption.

The phylogeny of known fossil lions, although subject to much controversy and disagreement among specialists, is now becoming clearer. As described above, the earliest known lion-sized *Panthera* fossils are from Laetoli in Tanzania, but the oldest undisputed fossils of *Panthera leo* (figure 4.13), subsequently referred to as the "savanna lion" throughout this book, are about 2 Ma and are from the Lake Turkana Basin of Kenya and Ethiopia, establishing this as the oldest known lion species. Younger savanna lion fossils are known from Olduvai Gorge in Tanzania (1.75 Ma); the French Mediterranean site of Vallonnet (900 ka); Isernia, Italy (700 ka); and Mauer, Germany (550 ka). *Panthera leo* survives today in sub-Saharan Africa and in the Gir Forest of India and probably shared a common ancestor with the steppe lion *P. spelaea* (see later in this chapter) no more than 200 ka and possibly as recent as 70 ka. Genetic studies demonstrate that North African and Asian subspecies of *P. leo* form a distinct lineage from sub-Saharan savanna lions and are phylogenetically basal among living populations. In spite of the closer geographic proximity of northern savanna lions with extinct steppe lions, the living sub-Saharan animals are genetically closer to *P. spelaea*, suggesting an early dispersal of a common ancestor before the separation of the northern *P. leo* stock. As noted in chapter 1, some specialists recognize the subgenus *Leo*, hence *Panthera (Leo) leo*.

The next oldest recognized lion species, **Panthera fossilis** (figure 4.14), is known from an early Pleistocene site, the Kuznetsk Basin of western Siberia (1.07–0.99 Ma), the Cromer Forest Bed of England (700 ka), and a number of later middle Pleistocene (500 ka) sites in western Europe. Often informally referred to as the "early Middle Pleistocene cave lion," *Panthera fossilis* and other "cave lions" are more correctly termed "steppe lions" because of their actual habitat. *P. fossilis* was longer at 2.4 m (7.9 ft.) than today's African lions, and probably weighed about 226 kg (500 lb.). Although there isn't evidence at this time, the possibility exists that during the mid-Pleistocene populations of *P. leo* may have dispersed northward into Eurasia along land corridors created by lowered sea levels during glacial periods, evolving into the much larger *P. fossilis*.

Some specialists maintain that all known fossil lions are subspecies of the extant savanna lion, *P. leo*, but detailed morphometric analyses by

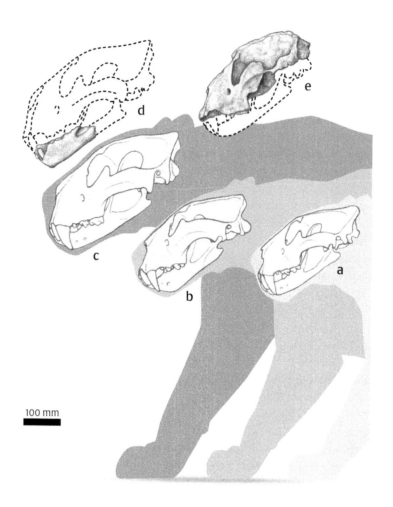

100 mm

FIGURE 4.13. Comparison of lion skulls.
The sequence from the extant savanna lion *Panthera leo* (a) through the steppe lion *Panthera spelaea* (b) to the American lion *Panthera atrox* (c) displays a longitudinal size gradient, or cline, that reflects an eastward migration from an African source. The ancestral form of the steppe lion, *P. fossilis* (d), was also larger than the extant lion. The giant Kenyan lion (e) has been attributed to the extant species but is comparable in size to the largest steppe lions. *Source*: Skull drawings modified from Saragusty et al. (2014); Sotnikova and Nikolskiy (2006); Manthi et al. (2017).

Marina Sotnikova and Pavel Nikolskiy, Sotnikova and Irina V. Foronova, and G. F. Baryshnikov (all of the Russian Academy of Sciences), as well as molecular studies by Ross Barnett (University of Copenhagen) and associates in 2009, now strongly suggest that the savanna lion *Panthera leo*, the steppe lion *Panthera spelaea*, and the "American lion" *Panthera atrox* are all separate species.

FIGURE 4.14. A pair of female *Panthera fossilis* steppe lions attacking a female gallic elk (moose), *Alces gallicus*, in a marsh.
During the early to mid-Pleistocene, fossils of large ungulates like elk and deer are known from the "Weyburn Crag" facies of the Cromer Forest Bed Formation of West Runton, England. These animals were probably common prey of the huge early steppe lion.

Panthera fossilis was the likely ancestor of the later *Panthera spelaea spelaea* (figure 4.15, plate 4 c, d), commonly called the "European" or "Eurasian" cave lion and known from a variety of Eurasian sites beginning at about 300 ka. *P. s. spelaea* is the iconic "cave lion" whose images were painted by Paleolithic artists on the walls of Chauvet Cave in southern France and sculpted as figurines by others at Vogelherd Cave in southwestern Germany (see chapter 5). Recently, Alain Argant and Jean-Philip Brugal (both Aix Marseille Université) have demonstrated, on the basis of morphometric studies drawn from a large database, the existence during the Middle Pleistocene of a distinctly smaller-sized population of *P. spelaea* from Igue des Rameaux in southwestern France. Erected as a new subspecies, **P. spelaea intermedia**, these lions may represent a size-based **cline** derived from the much larger, earlier *P. fossilis* to the smaller, later *P. spelaea spelaea*, or they may simply be a regional variant whose body dimensions could have been an adaptation to a smaller-sized prey base (see chapter 3).

Although its taxonomic status is uncertain, remains of a steppe lion described as **Panthera youngi** (or *P. spelaea youngi*) are found at some of the same levels as the hominin *Homo (Sinanthropus) erectus*, once known

FIGURE 4.15. Skull and life restoration of *Panthera spelaea spelaea*.
Known as the steppe lion or cave lion, *P. spelaea* was an iconic Pleistocene carnivoran.
It had a huge geographic range across Eurasia and competed for prey with hyenas,
homotheres, and human hunters throughout the mammoth steppes. As portrayed
in Aurignacian and Magdalenian cave art, males were apparently maneless.

as "Peking Man," at the Zhoukoudian (Choukoutien) Caves near Beijing
in northeast China. That cave system was occupied for an extremely long
period, between 700 and 200 ka, and the lion and human remains occur
in limestone fissures dating to about 350 ka. Smaller than *P. spelaea spelaea*,
and differing in skull proportions from other steppe lions, the subspecies
Panthera spelaea vereshchagini, the "East Siberian" or "Beringian" steppe lion
is based on finds occurring in Yakutia (Russia), Alaska (United States),
and Yukon Territory (Canada) that are all dated about 350 ka. Although
some specialists disagree with its subspecific status, *P. s. vereshchagini*, if
valid, would logically be part of a natural cline or continuous size range of
this species from the UK to Canada (figure 4.16).

The Late Pleistocene "American" lion, **Panthera atrox** (figure 4.17, plate 4e), whose morphology has led some workers to interpret it as a jaguar, was truly a giant pantherin, reaching overall body lengths of 3.50 m (11.5 ft.) and weights of 351 kg (774 lb.) in males, rivaled in size only by some specimens of *P. fossilis* from the Cromer II Forest Bed in England and the "giant lion" from Kenya. Geologically, it is a very recent species, first appearing in the Rancholabrean LMS of the Late Pleistocene about 195 ka, and some of its less advanced cranial and dental features, according to Sotnikova, Nikolskiy, and Foronova, could be explained by the retention of such characters from earlier steppe lions. It became a distinct taxon when its populations were cut off from those of Beringian steppe lions (*P. s. vereshchagini*) by the confluence of the Cordilleran (western) and Laurentide (eastern) ice sheets. Biometric analysis by H. Todd Wheeler (La Brea Tar Pits Museum) and George T. Jefferson (Colorado Desert District Stout Research Center) of well-preserved skeletal material from the famous La Brea Tar Pits (Rancho la Brea) site in Los Angeles and from Natural Trap Cave, Wyoming, show that the limbs of *P. atrox* were about 10 percent longer than those of *P. leo* relative to skull length. Increased limb length is an adaptation to cursoriality or swift pursuit and, as with *P. spelaea* and the savanna lion, it supports the interpretation that these cats were open-country hunters. The researchers hypothesize that, because available North American woodland habitats were already occupied by the saber-tooth *Smilodon*, ancestral American lions were forced to inhabit open habitats exclusively; as a result they developed limb proportions that enhanced their running ability for capturing prey. Both study sites were natural traps for *P. atrox*. The proportions of male and female bones, when corrected for preservation bias as to sex and age, plus the high degree of sexual dimorphism shown by both bones and teeth suggest something else: that American lions, like modern lions, lived in social groups or prides.

It's reasonable to hypothesize that early in their evolution lions developed a pride structure and group territoriality. If similar to that of today's lions, a pride would have been a core group of closely related, cooperatively hunting females that fostered one another's cubs, with coalitions of sibling or nonrelated males defending hunting territories and mating prerogatives. A communal structure would maximize the ability to kill large prey and to defend it against other predators, possibly wolves and homotherin sabertooths but especially hyenas. Just as importantly, it would maximize the successful rearing of cubs thanks to the protection of the young by the pride males. As with today's African lions and spotted hyenas, in their Eurasian Pleistocene equivalents there would have been an ongoing seesaw of mutual antagonism and attempts by one species to steal prey from the other, with strength of numbers, aggressiveness, and hunger determining the outcome. Although the European steppe sabertooth *Homotherium latidens* and Asian *H. crenatidens* and *H. ultimum*

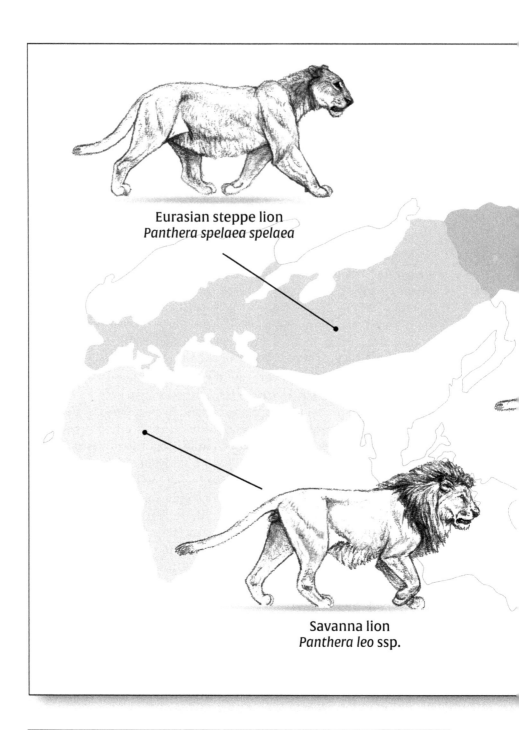

Eurasian steppe lion
Panthera spelaea spelaea

Savanna lion
Panthera leo ssp.

FIGURE 4.16. Late Pleistocene steppe lion, American lion and savanna lion ranges. During the time of the Last Glacial Maximum (about 20 ka) the combined range of Eurasian, Beringian, American, and savanna lion species was enormous, the largest of any known large, carnivoran cline. *Source*: Map modified from Barnett et al. (2009); Ersmark et al. (2015).

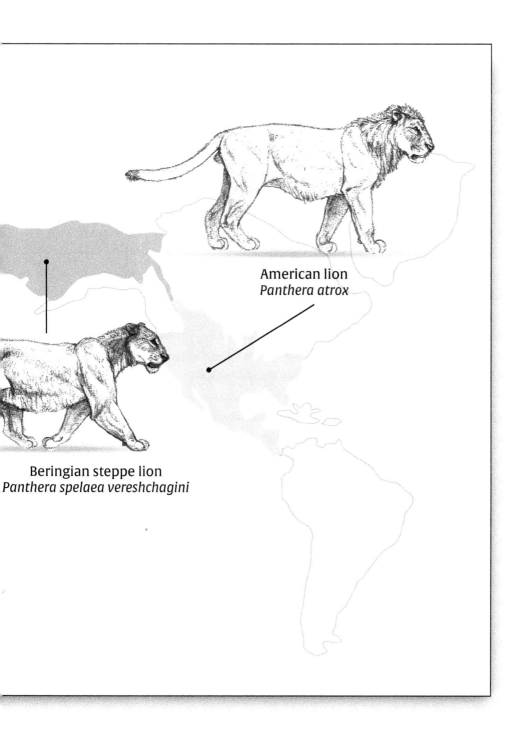

American lion
Panthera atrox

Beringian steppe lion
Panthera spelaea vereshchagini

FIGURE 4.17. Skull and life restoration of *Panthera atrox*.
Occurring only during the late Pleistocene of North America south of the two major ice sheets, the American lion may have been the largest cat ever to have existed. Its longer legs made it a faster runner than other lion species, and its huge size allowed it to take extremely large, fast-moving prey in open habitats. Whether *P. atrox* males possessed a mane like those of savanna lions is unknown, so its portrayal here is conjectural.

competed for similar prey with lions and hyenas, the more gracile fore-limbs and overall build of homotheres suggest that, like cheetahs, they would have risked injury during confrontations and would therefore have avoided conflicts with the other two predators (see chapter 6). Despite being less efficient than sabertooths at for making kills, steppe lion groups, like those of African lions, could under the right conditions have taken down juvenile and adult female mammoths and rhinos. The reason for territoriality vs. nomadic behavior in following migratory herds of prey is not so obvious. As Schaller has noted in the case of African lions, the tendency for females to sometimes scatter widely during individual hunts, unlike large canids that usually hunt as a group, would detract from maintaining the pride as an effective unit. Territoriality is also a valuable mechanism for restricting the number of individual animals that may hunt in a given area, thus preventing the depletion of prey.

The later Pliocene and earliest Pleistocene can be regarded as the zenith point in worldwide mammalian diversity. Africa today is home to the greatest number of mammalian species (as of 2013, there were 1,116 including five introduced species, i.e., a quarter of all known mammal species), and we look to the sweeping savannas of the continent's east for a glimpse as to what Earth was like before the end-Pleistocene megafaunal extinctions between 14.5 and 11 ka. However, the late Pliocene and early Pleistocene African faunas were matched in diversity by an amazing number of horned and hoofed forms in Asia and the other continents. These in turn supported a group of large predatory species as numerous and diverse as those in modern Africa (figure 4.18). During the late Pliocene Mazegouan (Villanyian) LMS in regions of Asia bordering the Tibetan Plateau, these included not only the metailurin (*Metailurus major*) and homotherin (*Homotherium hengduanshanense*) sabertooths and early pantherins (*Panthera* sp.) but also early cheetahs (*Sivapanthera* sp.); small felines (*Lynx shansius*, *Felis peii*); lightly built, cursorial hyenas (*Chasmaporthetes kani*); specialized bone-eating hyenas (*Pliocrocuta orientalis*, *Crocuta honanensis*); moderately predaceous bears (*Agriotherium* sp.); and the early dogs (*Eucyon davisi*, *Canis chihliensis*), among others.

FIGURE 4.18. Late Pliocene–Early Pleistocene north Asian carnivore guild. Assorted species typifying the carnivore guild of the Late Pliocene–Early Pleistocene Mazegouan-Nihowanian Asian LMS's. From left to right, the wolf *Canis etruscus*, the fox *Vulpes praeglacialis*, the cursorial hyena *Chasmoporthetes lunensis*, the bear *Agriotherium inexpectans*, the smilodontin sabertoothed cat *Megantereon nihowanensis*, the homotherin sabertoothed cat *Homotherium crenatidens*, the lynx *Lynx shansius*, and the marten *Martes pachygnatha*. The side of each square equals 1 meter (3.28 ft.).

Mammalogists and paleontologists use the term **guild** for a regional community of predators or herbivores, a term that's borrowed from the medieval and Renaissance concept of a group of professional artisans who made related but different craft goods to avoid directly competing with one another. In the case of predators, it implies an assemblage of localized carnivorans that are all predatory but prey on different elements of the herbivore community, thereby avoiding direct competition and maximizing their chances of success. In the Serengeti ecosystem of East Africa lions communally take down large game, sometimes the size of giraffes or even elephants, but preferentially prey on zebras or wildebeest. Stealth hunters like leopards ambush medium-sized antelope; cheetahs run down medium to small game through sheer speed; and spotted hyenas and Cape hunting dogs use stamina to relentlessly pursue and exhaust their prey. One measure of a vanished ecosystem's probable extent, stability, and duration over time lies in the numbers of **apex predator** species represented in a fossil assemblage. The presence of several apex predators (as with the Mazegoun Fauna of China) implies the existence of an even larger prey base comprising niche-partitioning herbivores—forming their own guilds and underneath the guild of predators like a foundation supporting the cap of a pyramid. It also implies that the apex predators were targeting different prey because if they were in direct competition for the same resources, the less well adapted predator species would lose their ability to compete and would become regionally or totally extinct (the principle of **competitive exclusion**). The evolving pantherins of the mid-Pliocene to later Pleistocene in northern Asia therefore had to play their own role in an already existing and diverse carnivoran guild with several apex predators, of which the machairodontines would have been the pantherins' closest competitors. Fossil evidence shows that both sabertooths and pantherins flourished and coexisted not only in Asia but other continents as well for about 200,000 years. Ultimately, however, the pantherins would challenge the continued dominance of the long-established sabertooths.

We know that the worldwide cooling trend, initiated in the Late Oligocene by the opening of the circum-Antarctic seaway, continued gradually but inexorably during the Miocene. The extensive tropical and temperate forests became replaced by more open grasslands, and herbivore guilds increased in diversity as a result. The forest fragmentation seems to have been enhanced toward the end of the Miocene with the spread of C_4 grasslands and, in Africa, resulted in the first appearance of the modern groups of mammals that exploited them in different ways. These included, among others, elephants, hippos, grazing pigs, and humans. In Eurasia as elsewhere, closed forest was still to be found but more open canopy woodlands, **riparian forest**, and savanna and edaphic (seasonally flooded) grasslands were the new norm. These open habitats were exploited by niche-partitioning herbivores, whose dietary specializations focused on different plants. In the diverse guilds of modern Africa, examples would be tragelaphin antelopes such as the bongo (*Tragelaphus eurycerus*) browsing in dense tropical rainforest; mixed-feeding impalas (*Aepyceros melampus* ssp.) at the forest edge; grazing reduncins like the waterbuck (*Kobus ellipsyprimnus*) and grazing bovins like the Cape buffalo (*Syncerus caffer*) in edaphic grasslands; and grazing alcelaphalins like the blue wildebeest (*Connochaetes taurinus mearnsi*) in the open savanna.

Antón, Turner, and other paleontologists have presented a convincing hypothesis describing the ecological relationships that affected the long-established machairodontines and the dispersing pantherins during the late Pliocene and Pleistocene periods. The closed to marginally open forest habitats in Eurasia and Africa, with characteristically adapted herbivore feeders, had, for millions of years, been well suited to the predation techniques of machairodontines, which would have used a short rush from close cover to ambush and overpower prey. With the fragmentation and reduction of the formerly extensive closed forests, this began to change (figure 4.19). While several genera of felines (*Felis*, *Lynx*, and others) had by late Miocene to early Pliocene times become ambush hunters of small prey such as rodents and birds, following the extinction of the nimravids the role of medium- to large-bodied stealth predators still largely belonged to the felid sabertooths. (Much later during the mid-Pleistocene, the medium-bodied felid predatory niche would be partially occupied by the pumas and their derived relatives the cheetahs; see chapter 1.). The sabertooths' formerly extensive woodland habitats became a mosaic of woodland and more open habitats that were invaded by early pantherin species like *Panthera palaeosinensis*. Competition with the machairodontines—large-bodied, closed-forest ambush predators—was inevitable. A combination of competition and increasing habitat restriction led first to a decline among the metailurin or short-toothed sabertooths, such as *Metailurus major* and *Dinofelis abeli*, that had the body proportions of forest felid predators and had been a diverse and successful machairodontine clade during much of the earlier Pliocene. By the Late Pliocene, metailurins were replaced by species like the dirk-toothed, early smilodontin sabercat *Megantereon cultridens* (figure 4.20, plate 8a). This was smaller than *Dinofelis* but had teeth and forelimbs well suited for preying on moderate-sized game such as woodland antelopes in Africa and deer in Eurasia. Woodland ruminants did not migrate seasonally and thus were available year-round to *Megantereon*, which was probably a solitary predator like living jaguars and tigers.

Some sabertooths, however, turned the tables on their adversaries. A second and very different line that arose in response to the increasingly open environments was the "scimitar-toothed" homotherin sabercat, *Homotherium latidens* (figure 4.21, plate 8b). Informally known as homotheres, these cats evolved as savanna and steppe predators with proportionately long forelimbs and modification of the paws for running and

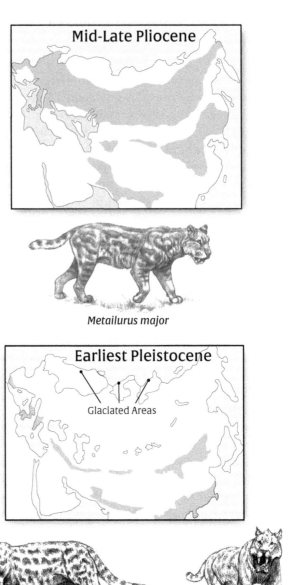

Metailurus major

Glaciated Areas

Panthera palaeosinensis   *Megantereon cultridens*

FIGURE 4.19. Comparison of mid- to late Pliocene and mid- to late Pleistocene closed forest.

Closed deciduous forest biomes predominated during most of the Pliocene and formed continuous bands (gray areas above) across Eurasia as late as the late Pliocene (3.6–2.58 Ma). These forests favored the predatory habits of "short-toothed" metailurin sabertooths like *Metailurus major* and *Dinofelis abeli*, which relied on dense cover to ambush prey. By the mid- to late Pleistocene (1.3 Ma–800 ka), however, increasing cold and aridity had led to a major reduction in closed deciduous forest. One sabertoothed lineage, *Megantereon*, survived and specialized as a predator of small to midsized ruminants in the remaining closed woodlands, but in open areas early pantherins such as *Panthera palaeosinensis* and its descendants replaced metalurins as dominant felid ambush predators.

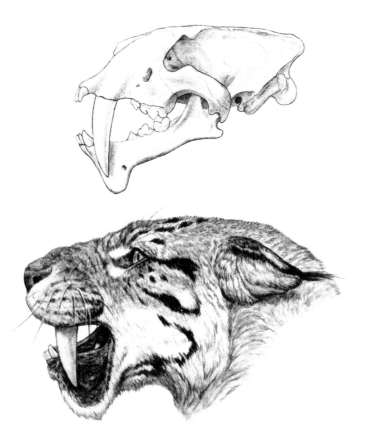

FIGURE 4.20. Skull and life restoration of *Megantereon cultridens.*
Megantereon is a smilodontin sabertoothed cat that was common in Africa, Eurasia, and North America during the Late Pliocene and Early Pleistocene. About the size of a large modern jaguar but heavier, like other machairodontines (except for homotherins) it was an ambush predator in open and closed forests. In North America it gave rise, during the Late Pliocene to *Smilodon gracilis,* probable ancestor of the larger, iconic *Smilodon fatalis.*

grappling. Although far more gracile in build than the previously discussed large, early steppe lion *Panthera fossilis* and than giant hyenas, *Homotherium* lived in social groups that would have made it competitive for hunting and defending kills. The long forelimbs and cheetah-like tractional ability of its paws gave this sabertooth an energy-efficient, hyena-like cantering gait that would have made it capable of covering long distances in following migratory prey. Having closed with its victim, the large dew claw and massive, projecting incisors helped to subdue an animal so that the upper canines could do their fast, effective killing. These adaptations made *Homotherium* a distinctive social predator and an effective competitor

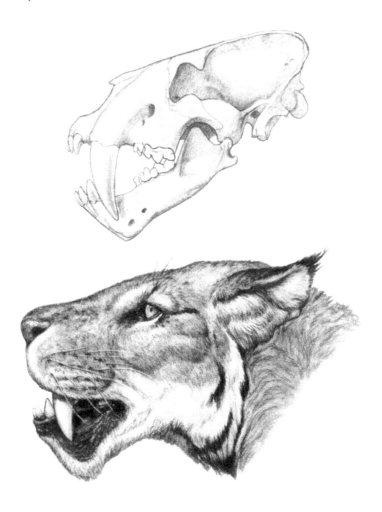

FIGURE 4.21. Skull and life restoration of *Homotherium latidens*.
Homotheres evolved different morphology and predation behavior from other
sabertoothed cats and were a successful form that competed with other open-country
carnivorans in Africa, Eurasia, and North and South America during the mid- to
late Pleistocene.

of early lions on the open plains, whereas in more forested habitats the
solitary *Megantereon* and *Hemimachairodus zwierzyckii* held their own against
the early pantherin newcomers that were now dispersing across Eurasia,
Africa, and North America.

As Agustí once eloquently wrote (Agustí and Antón 2002), the close
of the Pliocene was, in a real paleobiological sense, the end of a world.
The next epoch, the Pleistocene, ushered in rapidly changing climates
and altered ecosystems that had long been relatively stable. In some cases
this resulted in regional extinctions of major mammalian groups that had

been diverse and successful for millions of years. Although the number of species involved was relatively limited, in Eurasia they included both open-plains dwellers like hipparionine horses and forest forms such as deinotheres, gomphotheres, giraffids, and, as discussed, metailurin saber-tooths. Their role was replaced the Pleistocene by *Equus* horses, giant elk and deer, mammoths, early modern humans, and pantherins (plate 5). The stage was set for the next ice age.

CHAPTER 5
Testimony of the Caves

Zoolithenhöle cave, Germany, 58,000 years ago: The silence and darkness are profound, and the two female *Panthera spelaea* lions are guided only by the powerful, millennia-old stench of the cave bears' hibernation and maternity dens, which grows ever stronger as they creep through the passageway. The subadult moves unsteadily because of brain injuries from a hyena's massive bite to her head the year before when, at the point of starvation from her pride's poor success in hunting, she had tried stealing a deer leg from a hyena clan's food cache. Earlier that winter she and her pride sister had begun probing the innermost recesses of the caverns to kill and escape with a cub of the giant cave bear *Ursus spelaeus.* This highly dangerous practice had twice paid off, with each lion grabbing a cub and racing back along their path before the sleepy mother could respond. They are inside the maternity den when the rattle of a misstep on the limestone floor awakens the cub, whose sudden bawling arouses the mother bear. While both lions are large, the injuries of the first seals her fate: closer and slower to respond to the enraged mother bear than her pride sister, the lion receives the full fury of the massive paw to her head, which kills her instantly. The other lion flees. After a few growls and sniffs of the lifeless body, the mother ambles over to lick her frightened cub (figure 5.1).

. . . . .

Although it's easy to think of caves as silent, inert places, their ecosystems constantly change during the many phases of their existence. They are home to biological processes that may include permanent colonization by specialized plants, insects such as crickets, and, sometimes, blind shrimp and fish that are found only in its pools. More sporadic but often long-lasting use as roosting areas by bats and owls takes place. Sometimes the cave will become closed off and dormant, perhaps for thousands of years, until further erosion allows it once again to become active. Among the notable if sometimes temporary vertebrate inhabitants are carnivorans. Some cave systems are hundreds of thousands of years old and may have provided temporary shelters or permanent dens and food-storage areas for a succession of small and large predators. A series of caves spanning latitudes from South Africa to Romania have yielded thousands of pantherin fossils dating from the very earliest Pleistocene through the beginning of the Holocene epoch. Their stratigraphic sequences contain evidence of long-term climate changes and ecological successions but have occasionally documented big cats' first fateful encounters with humans.

FIGURE 5.1. *Panthera spelaea* females approaching a cave bear den.

On the southern slopes of the greater Caucasus Mountains overlooking the Djedori River, South Ossetia, lies an intriguing window into pantherin paleoecology. Discovered in 1955 by the paleontologist Vasili P. Liubin (Lioubine) (Zoological Institute of the Russian Academy of Sciences, St. Petersburg), who then excavated and studied it, the Kudaro Paleolithic Site is a complex of several eroded and exposed caves stacked one above the other and represents a remarkable series of intervals in Pleistocene time (figure 5.2). Beginning about 360 ka, the lowermost sediments of Kudaro 1 (brownish-yellow sandy clay/clayey soil) accumulated during the Holsteinian interglacial phase of the Middle Pleistocene, and they contain fossils from a variety of temperate-forest-dwelling mammals ranging from macaque (*Macaca* sp.) and porcupine (*Hystrix indica*) to Asian black bear (*Ursus thibetanus*). Appearing among these warmth-loving species are the bones not only of such small cats as the African/Eurasian wildcat (*Felis silvestris*) but also of the Eurasian jaguar *Panthera gombazoegensis*. The leopard (*Panthera pardus*) and wild cat are found at all levels, and the Eurasian cave lion *Panthera spelaea* and possibly the Pleistocene Eurasian lynx (*Lynx issiodorensis*) are from the most recent, yellow to dark-gray loams. The presence of the three pantherins within the cave indicates that by this time in the Middle Pleistocene the jaguar, leopard, and lion lineages had gone their separate ways. Fossil bones of felids and other carnivores are not unexpected in caves because these have always been used for shelter, for consuming prey, and as denning areas, but the question arises as to why *P. gombazoegensis* is known from only the lowermost Kudaro levels, *P. pardus* and *F. silvestris* from all levels, and *P. spelaea* (and possibly *L. issiodorensis*) mainly from the most recent ones. Judging from close living and fossil relatives, all of the felids known from Kudaro had some degree of cold tolerance if necessary, but the later disappearance of *P. gombazoegensis* and its replacement with the other large and small cats is mirrored by a corresponding sequential distribution of cave bears at Kudaro. *Ursus thibetanus* from Kudaro 1 was followed by the Late Pleistocene "Kudaro cave bear," *Ursus kudarensis*, and then by two cave bear species, *U. spelaeus* and *U. ingressus*. It would appear that Eurasian jaguars and Tibetan bears found at the earlier levels of Kudaro were less cold-tolerant than lions, leopards, and later cave bears. *P. spelaea* (and, if actually present, *L. issiodorensis*) are thought by Baryshnikov to have been relatively recent arrivals to the Caucasus region in the Late Pleistocene, whereas in the southern Caucasus leopard bones are known from many earlier localities, indicating that it was already a widespread species by this time.

The fall in temperature that accompanied the replacement of jaguars by leopards and lions at Kudaro resulted in habitat change. Palynological (pollen) evidence from the cave's stratigraphic sequence levels paints a picture of climate change similar to that at Dmanisi in the southern Caucasus ranges, where interglacial warm-temperate forests, together with some

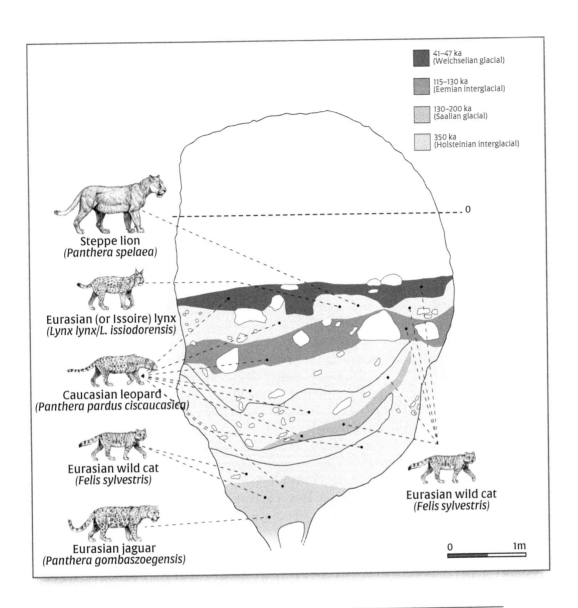

FIGURE 5.2. Feline succession at Kudaro caves 1 and 3, Greater Caucasus, Southern Ossetia.

A composite stratigraphic cross-section of Kudaro Caves 1 and 3 indicates that the Eurasian jaguar did not occur in the Caucasus region after the Holsteinian interglacial (350 ka) and that steppe lion and lynx did not appear until the Weichselian glacial (47 ka). Eurasian wildcat and Caucasian leopard are distributed throughout the sequence and persisted in the region until the Holocene. *Source:* Cave section after Baryshnikov (2011); animals to scale.

savanna, were replaced during glacial phases by cool-temperate, deciduous and boreal-conifer forest along with more extensive, cold-adapted steppe. If Eurasian fossil jaguars were like extant South American subspecies, they would have inhabited dense, closed gallery forests bordering major rivers, as well as some savanna, during the earlier warm pulses. Like Pleistocene leopards, Eurasian jaguars may also have preferred environments ranging from lightly wooded savanna to scrub-covered, montane areas. Lions, then as now likely to be social predators, probably occupied the open savannas that, with the onset of glacial pulses, became cold steppes but avoided closed deciduous and conifer forest. The latter were already occupied by the ancestral Amur tiger and leopard, established hunters of the northern taiga-conifer forest and open woodlands.

The principle of competitive exclusion, explained earlier, offers another insight into the pantherin succession at Kudaro and possibly other areas of western Asia during the Pleistocene. The similarity of coat patterns in jaguars and leopards and their similar body proportions, with relatively short legs and longer bodies, suggest a broadly similar style of hunting small- to medium-sized prey in closed- to open-forest conditions. This does not mean that coexistence was impossible as remains of both species occur at the same stratigraphic levels in two sites in France and the Czech Republic, but elsewhere the overlap in their prey preferences and similarity in hunting behavior may not have allowed them to share the same environment. In modern ecosystems like India's Kanha National Park, tigers actively kill leopards, as do lions in the Serengeti of Tanzania—not only because of competition for prey but also because the smaller cat is a threat to the larger's cubs. In Eurasia leopards may have ultimately won out as the predominant medium-sized pantherin hunter because of habits such as hauling their prey into trees to avoid theft by other predators, which jaguars and other bigger pantherins do not attempt.

The window to the past provided by Kudaro and other cave sites is one of several that show us glimpses of big cat evolution during the Pleistocene. As with the Pliocene, this epoch was characterized by major climate-ecosystem changes and faunal turnovers—creating new habitats and prey opportunities for pantherins. Following an initial dry, cold period (the Asian Nihewanian or European Eburonian LMS), the early Pleistocene began to alternate, on a general basis of 100,000-year cycles beginning at 1.7–1.8 Ma, between glacial intervals of major cool to extreme stretches of cold, generally lasting 80,000 years, and 20,000-year-long warm interglacials, sometimes actually warmer than today's temperatures. Glacials were in turn punctuated by relatively brief warm pulses, **interstadials**, and the interglacials, by cool ones, **stadials** (figure 5.3). Besides locking up trillions of tons of Earth's water into vast ice sheets, lowering sea levels by an average of 91.44 m (300 ft.), and exposing huge new land areas, glacial periods had other profound effects on northern-latitude ecosystems. One was that

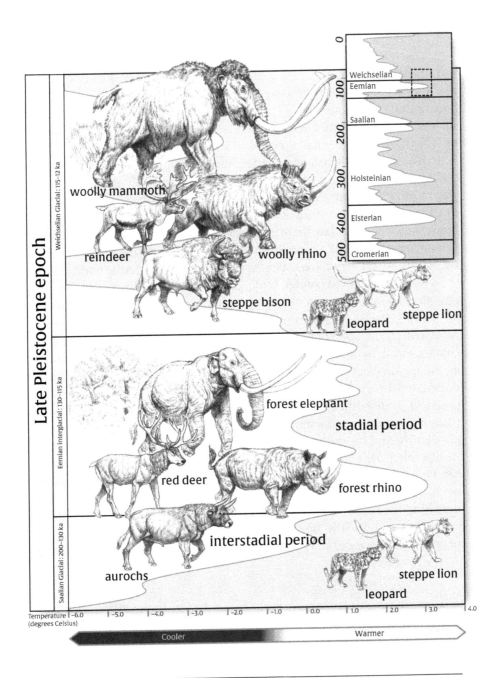

FIGURE 5.3. Succession of glacials and interglacials and faunal turnover. The later Pleistocene of Eurasia, as in North America, comprised a sequence of alternating cool-to-cold glacials and mild-to-warm interglacials (see inset), each characterized by distinctive biomes and mammals. In northern Europe the Eemian interglacial featured mixed oak-hornbeam deciduous forests inhabited by straight-tusked or forest elephant (*Palaeoloxodon*), red deer (*Cervus*), woodland rhino (*Stephanorhinus*), and aurochs (*Bos*), among others. The succeeding Weichselian glaciation was dominated by steppe grassland where woolly mammoth (*Mammuthus*), woolly rhino (*Coelodonta*), reindeer (*Rangifer*), and steppe bison (*Bison*) thrived. The same pantherins successfully took advantage of available prey during both climatic extremes. *Source:* Glacial cycles modified from Imbrie et al. (1993). Faunal successions based on Kurtén (1968); Deidrich (2014); and Baryshnikov (2016).

in Eurasia, as in North America, to an even greater degree than during the Miocene and latest Pliocene, large regions of forest were replaced during the Pleistocene by cool steppes dominated by grasses, forbs, and shrubs. In the most northerly areas a new type of ecosystem, **tundra**—a usually treeless biome characterized by some grasses, sedges, mosses, and lichens—came into existence. During the cold to cool glacial and interstadial periods ice-sheet expansions caused the normally extensive, temperate broadleaf- and mixed-forest zones to shift south latitudinally and to contract in most regions. Conversely, during the warmer and wetter interglacials and stadials these expanded to again become more varied, producing a mosaic of alpine flora, lightly forested tundra, Siberian conifer forest, mixed forest (oak, beech, hornbeam, other hardwoods), taiga or boreal forest (mixed conifer, birch, hardwoods), and rivers with riparian or gallery forests. (figure 5.4).

During glacial periods Eurasia, although it featured highly diverse ecosystems that spread latitudinally from west to east, was especially characterized by long swaths of "open" **biotopes** or ecosystems known as "steppe-savanna" or "mammoth steppe" that contained a diverse flora dominated by grasses. These habitats supported herbivore guilds composed of large-bodied, mainly grazing or mixed-feeder species such as mammoth, horse, woolly rhino, bison, giant deer, saiga antelope, and others that thrived on the nutritious mix of short to tall grasses, forbs, and shrubs (plate 5). The geographical origins of these cold-adapted mammals can be traced to the Tibetan heartland region, from which they had dispersed during the latest Pliocene and earliest Pleistocene, readily accommodating themselves to the cold temperatures that came to increasingly prevail throughout the vast continent's northern latitudes. Contrary to popular belief, ice-age ecosystems were not subjected to constant frigidity.

FIGURE 5.4. Comparison of interglacial and glacial biomes.
During the apex of the Eemian interglacial (~100 ka) Eurasia was a mosaic of predominantly forested regions that supported a variety of warmth-adapted mammals, typified by forest (straight-tusked) elephant (*Palaeoloxodon*), red deer (*Cervus*), horse (*Equus*), Merck's rhino (*Stephanorhinus*), aurochs (*Bos*), dirk-toothed sabertooth cat (*Megantereon*), Eurasian jaguar (*Panthera gombaszoegensis*), steppe lion (*Panthera spelaea*), hyena (*Crocuta*), leopard (*Panthera pardus*), tiger (*Panthera tigris*), and Neanderthal humans (*Homo neanderthalensis*). By the Würm glacial's maximum cold peak (~20 ka), however, most forest areas had long given way to large areas of dry, cold steppe. Here cold-adapted forms like woolly mammoth (*Mammuthus*), reindeer (*Rangifer*), woolly rhino (*Coelodonta*), steppe bison (*Bison*), homotherin sabertooth cat (*Homotherium*), steppe lion, leopard, tiger, and Cro-Magnon or early European modern human (*Homo sapiens*) gradually replaced the earlier species.
Source: Geography and vegetational zones modified from Kurtén (1968); Velichko, Andreev, and Klimanov (1997); and Deidrich (2013).

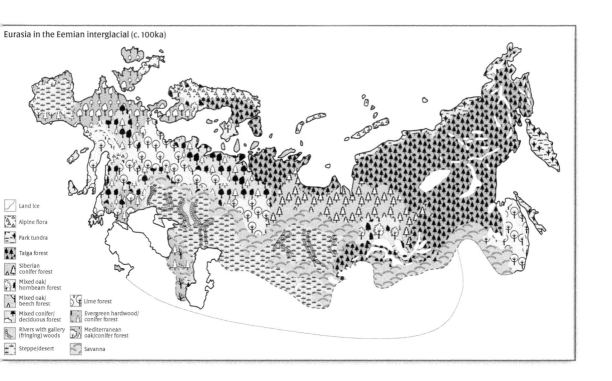

Eurasia in the Eemian interglacial (c. 100ka)

Land Ice	
Alpine flora	
Park tundra	
Taiga forest	
Siberian conifer forest	
Mixed oak/hornbeam forest	
Mixed oak/beech forest	Lime forest
Mixed conifer/deciduous forest	Evergreen hardwood/conifer forest
Rivers with gallery (fringing) woods	Mediterranean oak/conifer forest
Steppe/desert	Savanna

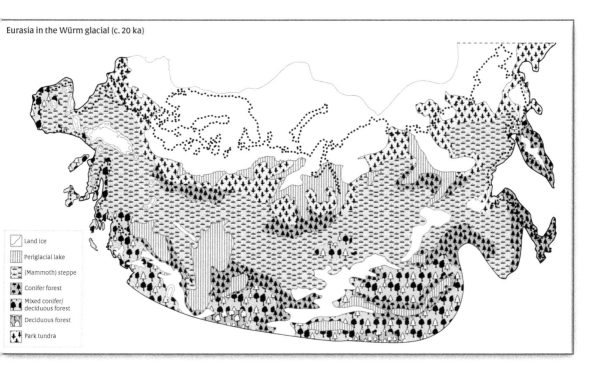

Eurasia in the Würm glacial (c. 20 ka)

Land ice
Periglacial lake
(Mammoth) steppe
Conifer forest
Mixed conifer/deciduous forest
Deciduous forest
Park tundra

Computer-generated models of glacially dominated weather patterns indicate a less extreme seasonal weather pattern than today's, moderating the excessive summer heat and bitter winter cold typical of today's inland continental regions. These high-latitude northern biotopes in many ways faunally mirrored the modern lower latitude, tropical to subtropical faunas in which elephants, zebra, and bovids predominate. The Eurasian steppe herbivore guilds in turn supported a rich array of predators dominated, as in Africa, by lions, homotherin sabertooths, and hyenas, with wolves, dholes, wolverines, and other northern carnivorans substituting for southern cape hunting dogs, jackals, and honey badgers. With the advent of warmer, wetter, more humid interglacials and stadials, steppe biotopes contracted and retreated north to give way to coniferous and mixed temperate broadleaf/conifer forests that favored warmth-adapted faunas. During interglacials such as the Eemian (100 ka) in Britain, woodland-dwelling aurochsen replaced their steppe-bison cousins, and hippos basked on the banks of the River Thames. These and the "forest" or "straight-tusked" elephant (*Palaeoloxodon antiquus*) and Merck's rhino (*Stephanorhinus kirchbergensis*) would migrate south when cooler conditions returned. Meanwhile, the cold-adapted forms that previously inhabited the region had migrated northward with the habitats they needed and remained there until conditions shifted once more toward a glacial. Such shifting environmental **refugia** usually made survival possible for most species during the extreme climatic fluctuations of the Pleistocene ice age.

Against this changing backdrop several mammalian dispersals took place. In northern, central, and eastern Asia during the very early Pleistocene Asian Nihewanian LMS, several clades of medium-large (horses, deer) to very large ungulates (mammoths) radiated from their evolutionary centers of origin to elsewhere in Africa, Eurasia, and North America; as a result some of these new, larger prey species offered pantherin cats a different and broader prey base. By the unfolding of the first major glacial period, the modern horse genus *Equus* had dispersed from the New World across the Beringian land connection, to give rise to such Asian species as *Equus sanmeniensis*. The migrating horses were accompanied by the camel *Paracamelus gigas*, and by the early to mid-Pleistocene native Eurasian ungulates came to include the Asian giant deer *Sinomegaceros konwanliensis*, early representatives of sika and axis deer like *Cervus (Sika) grayi* and *Axis shansius*, the antelopins *Antilospira licenti* and *Spirocerus wongi*, the early bovins *Leptobos brevicornis* and *Bison palaeosinsensis*, and the Eurasian wild pig *Sus lydekkeri. Equus* mingled with the last Asian three-toed horses (*Hipparion sinense*), and other genera began to give a modern aspect to the Asian ungulate community. Some Asian ruminants, in turn, dispersed westward into Europe, among them the rupicaprin tahr *Hemitragus albus*, ancestor of the modern Himalayan tahr *H. jemlahicus*.

FIGURE 5.5. North Asian pantherins, machairodontines, and prey. During the Middle Pleistocene of northern Asia, pantherins came to share with machairodontines the roles of apex felid predators in both open steppe (top row) and forest habitats (bottom row) and preyed on a variety of herbivores characteristic of each environment and climactic period. Top row, from left to right: the steppe lion *Panthera spelaea* and the homotherin sabertooth *Homotherium crenatidens* and some of their likely prey species: the horse *Equus sammeniensis*, the reindeer *Rangifer tarandus*, and a *Mammuthus trogontherii* calf; although probably not a common prey item, mammoth calves were occasionally the cats' victims. Bottom row, from left to right: the Eurasian jaguar *Panthera gombaszoegensis*, the tiger *Panthera tigris*, and the smilodontin sabertooth *Megantereon nihowanensis* and their likely prey species: the large deer *Eucladoceros boulei*, the pig *Sus* cf. *S. lydekkeri*, and the muntjac, *Muntiacus bohlini*. The side of each square equals 1 meter (3.28 ft.).

Depending on the stadial or interstadial that was prevalent during the early Pleistocene and the type of region, ancestral Chinese tigers (*Panthera zdanskyi*) and leopards (*P. pardus* ssp.) probably shared with the sabertoothed cat *Megantereon nihowanensis* (and, during the Holsteinian interglacial, with the Eurasian jaguar *P. gombaszoegensis*) a prey base that consisted of midsized to smaller deer (*Eucladocerus boulei*), the muntjac (*Muntiacus bohlini*), pig (*Sus* cf. *S. lydekkeri*), and other closed-forest and open-woodland herbivores (figure 5.5). Elsewhere, prides of the steppe lion *P. fossilis* hunted migratory, open-country forms like the camel, horse, and bison as well as large sheep like *Megalovis latifrons* and the "giant" muskox *Praeovibos priscus* and its contemporary, extant muskox relative, *Ovibos moschatus*. For these they directly competed with the scimitar-toothed homotherin *Homotherium cranatidens*, and both felids would have occasionally taken the young of ancestral woolly rhinos (*Coelodonta nihowanensis*), early interglacial proboscideans like the forest elephant *Palaeoloxodon* (cf. *P. namadicus*), and early glacial forms like the enormous steppe mammoth, *Mammuthus trogontherii*.

Ancestors of the modern snow leopard *Panthera uncia* would have preferred high-altitude habitats, probably overlapping to some extent at lower elevations the preferred habitats of the ancestral common leopard, *P. pardus*. As the earliest known pantherins that evolved as predators of the high, steep Himalayan terrains, the ancestors of modern snow leopards in their more extended northern Pleistocene range continued to take rupicaprins similar to the large European goral (*Gallogoral meneghinii*) and the extant chamois (*Rupicapra rupicapra*), as well as now derived caprins like bharal, the Himalayan ibex (*Capra siberica*), and the sheep (*Ovis shantungensis*).

Yawning up at the sky in what is now the Cradle of Humankind World Heritage Site, near Sterkfontein in South Africa, Swartkrans is the geologically second-youngest of a group of five caves that includes Sterkfontein, Kromdraai, Makapansgat, and Taung. All are located northeast of Johannesburg in the Transvaal region's high **veld**. Known as karst sinkholes, these caves formed by the action of ground water that seeped for thousands of years through the dolomite limestone bedrock, slowly dissolving it to create steep vertical shafts and chambers far below ground level. Karst openings occur when upper rock dissolution finally causes the cave roof to collapse, forming a hole at the surface. A steep-sided sinkhole provides a hazard to any animals that fall in and can't climb out, dooming them to a slow death from thirst and starvation. Over time an accumulated cone of washed-in soil, debris, and animal remains builds up, preserving layers that include pollen and fossil bones that document distinct paleoecological and climatic sequences. With deposits dating back to the early Pleistocene (1.8–2 Ma), Swartkrans preserves thousands of mammal remains from species present when the surrounding area was an open savanna. A huge number of the bones belong to baboons, but some 140,000 are from our relatives—the robust australopithecine *Paranthropus robustus* and early members of our own genus, *Homo erectus* and *Homo habilis*.

The outer surface of one juvenile *Paranthropus robustus* cranium shows two neat holes toward the rear of the skull. These were interpreted by earlier paleoanthropologists to be wounds from a sharpened-bone or tooth-studded hand-held weapon that were inflicted during an attack by another australopithecine. In cataloguing this and other specimens, however, Charles K. "Bob" Brain (Transvaal Museum) noted that baboon skulls sometimes also showed the same pattern of twin punctures at the same location on the skull. This becomes significant because at Swartkrans and other caves, trees that have difficulty getting started elsewhere on the veld are able to prosper and grow tall at the lip of the sinkhole, their roots taking hold in the nutritious soil of the cave and the moisture washed in by the annual rains. Then as now, trees made fine places for leopards to hoist their kills out of the reach of other predators such as hyenas, lions, and sabertooths. These victims were carried into the branches overgrowing the sinkhole mouth, and any unconsumed body parts eventually fell down to join the ever-growing talus cone of bones and other debris. An efficient way for a leopard or other big cat to drag a large-skulled primate to a safe place is to hold it securely by the head, with the cat gaping its jaws so that the upper canines securely hook under the upper rim of the orbits while the lower ones penetrate the rear upper skull. That this is what produced the characteristic punctures on the hominin and baboon crania can be confirmed by holding some punctured skulls of baboons or hominins next to a mandible of an average-sized adult leopard fossil found at the same site; the spacing of the apices of the lower canines is a perfect match to the holes in the *A. robustus* cranium (figure 5.6).

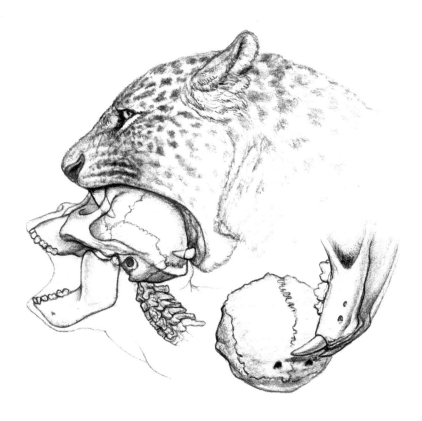

FIGURE 5.6. Leopard tooth marks on hominin skull.
The matching holes at the rear of some fossil baboon and hominin skulls from the cave site of Swartkrans, South Africa, are explained by the habit of ancient leopards of carrying their large-headed prey away from the kill site for consumption in a sheltered area. *Source:* Modified from an original concept and drawing by Jay H. Matternes (1985).

In addition to carrying their prey to more accessible caves so they could eat in seclusion, the ancestral common leopard *Panthera pardus*, *Dinofelis piveteaui*, and *Homotherium ethiopicum* (or *hadarensis*) may have also found caves to be worthwhile hunting grounds. In his studies Brain noticed the extremely high ratio of similar-sized and juvenile primate bones—baboons, monkeys, and hominins—to those of other mammals. He speculated that these fossils represented species that were sheltering in the caves during the chilly autumn and spring nights on the high veld and conducted an impromptu experiment to confirm his hunch. Arriving early in the evening at Uitkomst Cave, he hid on a raised ledge inside while a troop of about thirty baboons entered and gradually settled in for the night. After they were settled, he emerged, sending the baboons into terrified confusion, but, in spite of their fear, none could be made to leave the cave. From this experiment Brain concluded that the aroused animals, more afraid of the dangers possibly awaiting them outside, voluntarily chose to stay where at least a few would have been easy prey for the cats. That the majority of primate fossils in the deposit represented prey rather than natural deaths was confirmed by the paleoecologist Elisabeth S. Vrba (Yale University). She demonstrated that leopards or other felids are responsible for bone accumulations made up of prey species of about the same size and consisting of many young animals, whereas accumulations of prey remains varying greatly in body size are produced by mammalian scavengers such as hyenas. Despite the potential risk, that some Plio–Pleistocene hominins entered the Transvaal caves voluntarily, like the modern baboons, is supported by the fact that stone tools as well as their bones were found in some deposits. The lack of charcoal in these levels confirms that the use of fire was unknown at the time and so could not be used to help keep predators away.

For later pantherins and other carnivorans caves could be both hunting grounds and death traps, and this chapter's opening vignette, although dramatic, is closely based on known facts. Cajus G. Deidrich (PaleoLogic Research Institute), a specialist in the paleoecology of Middle to Late Pleistocene European fauna, has demonstrated the remarkable interactions of two top predators, the steppe lion (*Panthera spelaea*) and the now regionally extinct Eurasian spotted hyena (*Crocuta crocuta spelaea*) with two nonsympatric species of herbivorous cave bears, *Ursus spelaea* and *U. ingressus*. The interactions of these carnivorans are documented mostly from caves of southern Germany, the Czech Republic, and Romania, though there are also some open-air sites and important cave localities in the United Kingdom and France. One of the earliest to be excavated, Zoolithenhöle cave in Upper Franconia (Bavaria), Germany, is also one of the largest systems and records the life-and-death competition for survival among these animals.

Because its bones were first discovered in the fossil cave bear den of cave d'Azé, France, and others during the early nineteenth century, the name "cave lion" has often been applied to the mid–late Pleistocene *Panthera spelaea* and by extension to its early–mid-Pleistocene presumed ancestor, *P. fossilis.* Like modern African and Asian lions, however, these were predators of the steppes and marginal areas near forests, and so the name "steppe lion" is used in this book. Nevertheless, at least *P. spelaea* may sometimes have used caves as either temporary or permanent dens for shelter or raising cubs, since young cub material is known from Mixnitz, Aufhauser, and Hohlenstein caves, although these could also be the remains of hyena kills. (That tigers closely resembling the modern Amur tiger [*Panthera tigris altaica*] used caves at least occasionally as dens during the late Pleistocene is documented by the discovery in 1971 of young cub jaw fragments and deciduous ["milk"] teeth in Geographical Society Cave and others, in the valley of Partizanskya River in the Primorskii Territory of the Russian Far East.) The earlier *Panthera fossilis* was a huge pantherin that far exceeded the African lion in size and approached the size of the North American *P. atrox,* the largest known cat. Both *P. fossilis* and its probable descendant *P. spelaea* were of suitable size for killing, during interglacials, large open-forest bovids like the aurochs (*Bos primigenius*), giant deer (*Megaloceros verticornis*), "woodland bison" (*Bison shoetensacki*), and the Eurasian wild pig (*Sus scrofa*). Their prey also included open-terrain species like "steppe bison" (*Bison priscus*), reindeer or caribou (*Rangifer tarandus*), and European wild horse or tarpan (*Equus ferus*) during glacial pulses (plate 5).

Some of the steppe lion prey species have been positively confirmed by molecular analysis of *P. spelaea* bones both by Hervé Bocherens (Tubingen University) and his associates in 2011 as well as in 2015 by L. V. Kirillova (Ice Age Museum) and her colleagues. Intact fossil bone may sometimes preserve collagen, a structural protein abundant in this tissue and many others that acts as a sort of "glue" that binds structures together. The isotopic "signature" of herbivore proteins in the molecular structure of surviving collagen in the lion bone samples indicates that steppe lions were regional in their prey preferences. Reindeer was a favored prey of western Eurasian steppe lions (*P. s. spelaea*) during glacial periods, whereas eastern Beringian steppe lions (*P. s. vereshchagini*) in the area of Yakutia preyed heavily on horse and steppe bison (*Bison priscus*). Like modern *Panthera leo* and many other modern carnivorans, however, both were opportunistic, and *P. s. vereshchagini* sometimes took sheep and muskox as well as the occasional young woolly rhino and mammoth. Steppe lions in western Europe also sometimes targeted prey other than large ungulates. The young of the cave bear *Ursus spelaus* and the later *U. ingressus* were also on the menu, but not adults, a distinction made possible from the isotopic differences

FIGURE 5.7. *Panthera spelaea* pride attacking adult cave bear, *Ursus spelaeus.* When surrounded and caught in the open while foraging, solitary adult cave bears, especially if old or infirm, were probably vulnerable to predation by a pride of steppe lions.

between the bones of adult cave bears and those of their lactating cubs. This sample is not proof that steppe lions never attacked cave bear adults, however, since a pride could have certainly overpowered and killed an old, injured, or diseased solitary adult (figure 5.7) if they had surprised it in the open, as even solitary Amur tigers sometimes do today with Asiatic black bears (*Ursus thibetanus*) and Eurasian brown bears (*Ursus arctos beringianus, U. a. collaris*).

Evidence of steppe lion predation in high mountain areas on hibernating cave bear cubs—and possibly adults—occurs in Medvedia cave in Západné Tatry Mountains, northern Slovakia. Martin Sabol, Juraj Gullár, and Ján Horvát (Comenius University) documented the skeletal remains of four adult steppe lions, along with bones of adult and cub cave bears (*Ursus spelaeus*). During the interval preceding the Last Glacial Maximum (c. 20 ka) this relatively high elevation (1,133 m/3,717 ft.) was probably a closed forest zone dominated by conifers. It was an atypical habitat for open country, social predators like lions but the scarcity of local ungulates on the steppes during winter may have left the big cats with few other options.

Scarcity of prey may have had an effect on the hunting and other social behavior of *P. spelaea* as well. From the isotopic analyses and from biometric analysis of male/female skeletal ratios from Medvedia and other sites, Sabol and some other workers have proposed that rather than living and hunting in prides like savanna lions, steppe lions could have had a more solitary lifestyle. This is supported by the depictions of lions in the Chauvet cave paintings in southern France, which show hunting lions as individuals or as male and female partners, as well as in prides. Turner and Antón (1997), however, argue that steppe lions, because of their strong sexual dimorphism (in which males were typically 25 percent larger than females) probably occurred in prides, in which the huge size of the males helped to maintain territorial and breeding dominance. Guthrie makes a case for steppe lion prides having a maximum of three adult individuals per pride—a concept supported by the work of K. G. Van Orsdol (Uganda Institute of Ecology) and his associates in a study of the ecology of modern savanna lions. *P. spelaea*'s behavior may have resembled their southern counterparts, in which pride numbers simply decrease or increase in proportion to prey availability. At this point, we simply don't know.

Unlike the similar but inaccurate term for lions, the popular name "cave bear" for successive species like *Ursus deningeri, U. spelaeus* (plate 9d), and *U. ingressus* is fitting for an animal that, although a committed herbivore of the open steppes, river valleys, and mountain margins, spent the winter months of its life either in deep sleep or actual hibernation in caves and other protected areas. In large cave systems such as Zoolithenhöle, intact skeletons ranging from fetuses to old adults show that cave bears habitually and probably instinctively sought out the deepest cave recesses for their hibernation and maternity dens, primarily to protect their cubs against the predations of hyenas and lions (at first, *Panthera fossilis* and, during the later Pleistocene, *P. spelaea*). Hunting bear cubs in winter dens was risky but may have become more common when the steppe lions' typical prey grew scarce during stadials (see chapter 6). When they are found, the skulls of adult cave bears from cave areas outside hibernation and maternity dens sometimes have partially healed, bite-trauma damage to the upper surface of the posterior area of either right or left frontal bones, as do those of hyenas and steppe lions (figure 5.8). Large living carnivorans, especially felids, typically try to bite the head of another animal during fights, and this may represent an injury the lion inflicted on an adult bear during an attempted cub predation. The fact that these skulls and other associated bones occur farther toward (but outside) the denning areas suggested to Deidrich that in some caves there was a "conflict zone" in which bear fights with intruding lions were likely to

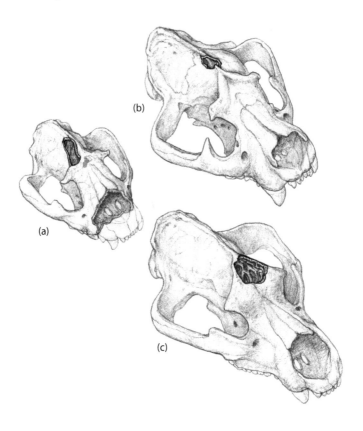

FIGURE 5.8. Bite injuries to skulls of European cave carnivorans.
Some skulls of (a) spotted hyena (*Crocuta crocuta spelaea*), (b) steppe lion (*Panthera spelaea*), and (c) cave bear (*Ursus spelaea*) from European mid- to late Pleistocene caves show partially healed damage to the upper cranium. These occurred from bites during lion-bear and lion-hyena conflicts.

happen (figure 5.9). A similar conflict zone, but closer to the cave entrance, involved lions and hyenas (see later in this chapter).

That lions were sometimes killed by adult cave bears when attempting this kind of predation has occasionally been revealed by fragmented or sometimes articulated lion skeletons found within the deepest accessible areas of caves. The earliest clear evidence of this comes from the Peştera Urşilor cave in Romania followed by that from Sloup, Výpustek, and Srbsko Chlum-Komin caves, Czech Republic, in which the most intact lion remains were found near the very back. The skeleton from Srbsko Chlum-Komin cave is currently the only articulated, virtually complete skeleton known of an adult female steppe lion. Exhibiting braincase bite trauma, it occurs near an *Ursus spelaeus* hibernation/maternity denning area, known from adult and juvenile cave bear bones and nests, toward

(a)

(b)

PLATE 1. Early ancestors of felids.
Typified by *Miacis incognitus* (*above*), whose appearance and habits may have been similar to a pine martin, miacids are the first known carnivorans. The fossa-like *Proailurus lemanensis* (*below*) is considered to be either close to or at the very beginning of the cat lineage.

(a)

(b)

PLATE 2. The discovery of *Panthera blytheae*, the earliest known pantherin. (*above*) Xiaoming Wang and his team in the Zanda Formation of Tibet. (*below*) The uncovered cranium of *Panthera blytheae* shortly after its discovery by Juan Liu. *Source:* Zanda Formation team photo courtesy of Xiaoming Wang; cranium photo courtesy of Jack Tseng.

PLATE 3. *Panthera blytheae* hunting among the hills of what would be known one day as the Zanda Formation, Tibet.

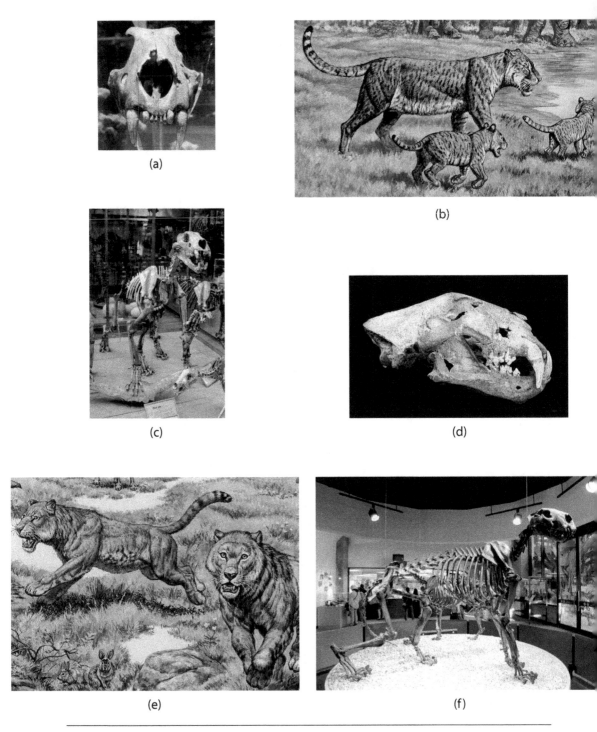

PLATE 4. Pleistocene pantherins.
(a) Skull of early modern tiger, *Panthera tigris acutidens*. (b) Restoration of early modern tiger, *Panthera tigris acutidens*, in the Amur River locality, northeastern Russia. (c) Skeletal mount of the "Lion d'Herme" (steppe lion, *Panthera spelaea*) skeleton at the Musee d'Histoire Naturelle, Paris. (d) Skull of steppe lion, *Panthera spelaea*. (e) Restoration of steppe lion, *Panthera spelaea*. (f) Skeletal mount of American lion, *Panthera atrox*, La Brea Tar Pits and Museum. *Source:* Photos (a, c, d) courtesy of Christine Argot, Musee d' Histoire Naturelle, Paris; photo (f) by Turi Hallett, Dryaduir Design.

(a)

(b)

PLATE 5. European Ionian steppe mammoth fauna (a) and Sino-Russian Zhoukoudianian fauna (b).

(a)

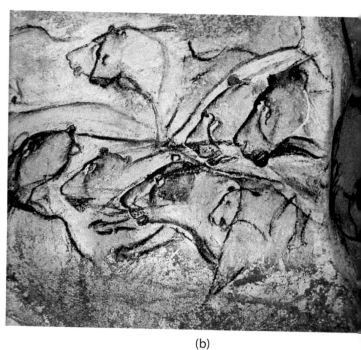

(b)

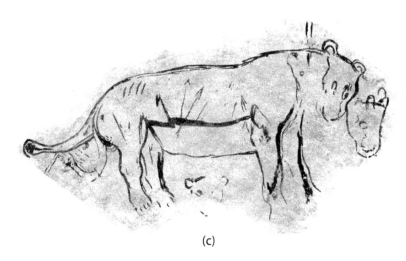

(c)

PLATE 6. Depictions of pantherins in caves.
Spanning tens of thousands of years, the art of the Cro-Magnon cultures has left us a rich assortment of big cat images, ranging from leopards through steppe lions as drawings and murals on cave walls. (a) The "Red Leopard" mural figure from Chauvet Cave, Ardèche, France. (b) The main frieze from the "Gallery of Lions" mural of Chauvet Cave, France. (c) Drawing of a lion (interpretive version by Abbe Breuil) from Cave of the Trois-Frères, Ariège, France. *Source:* Photos courtesy of the Musee d' Histoire Naturelle, Paris.

(a)

(b)

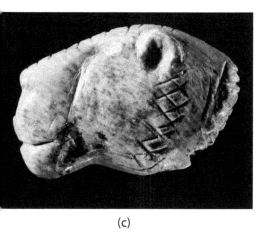

(c)

(d)

PLATE 7. Ice-age portable art of pantherins.
(a–c) Many small ivory carvings of pantherins and other animals, some carefully inscribed with interpretive patterns, are known from a 1931 discovery in Vogelherd Cave in the southwestern Swabian Jura, Germany, as well as from Dolní Vestonice, Czech Republic, and caves in France. (d) A unique Aurignacian age "therianthrope" sculpture, the "Lion-Man," known from the Hohlenstein-Stadel cave, Germany. *Source:* Photos (a) by Hilde Jensen, (b) by Don Hitchcock, (c) by Müller-Beck, (d) by Ralph Franken. Courtesy of Universität Tübingen.

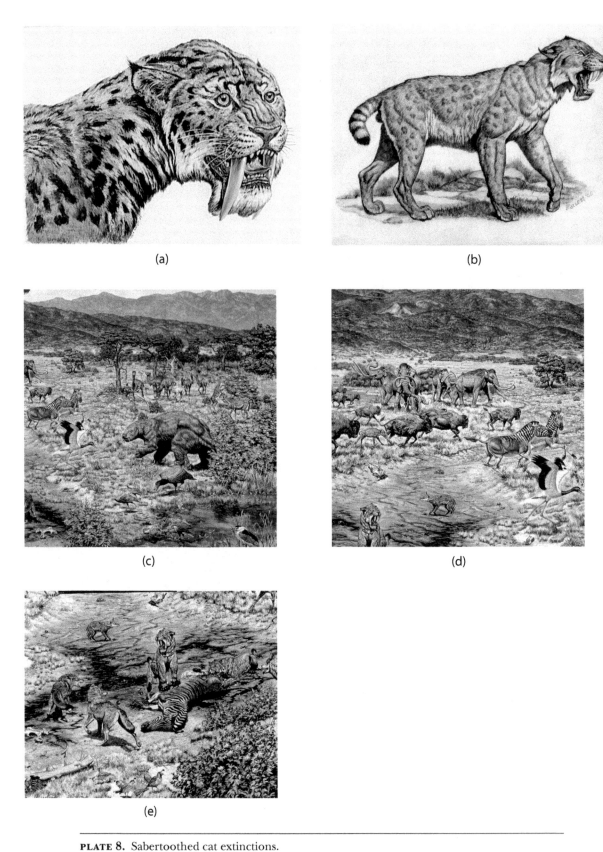

PLATE 8. Sabertoothed cat extinctions.
The end of the Pleistocene was a time of significant large-mammal extinctions, which in Africa, Europe, and Asia included the dirk- and scimitar-toothed sabertooths: (a) *Megantereon*, (b) *Homotherium*, (c, d, e) the animals of the Americas, however, typified by the Rancholabrean Fauna, remained diverse until about 12,000 years ago.

(a)

(b)

(c)

(d)

PLATE 9. Steppe and American lion extinctions.
Although the reasons for the disappearance of both steppe and American lions are still not completely known, a contributing factor may have been the climate-caused transition of the mammoth steppe in Eurasia and the American steppe, which in turn affected their prey. This consisted of mostly large ungulates. Some prey species are pictured: (a) bison, *Bison priscus*; horse, (b) *Equus ferus*; (c) reindeer, *Rangifer tarandus*; (d) cave bear, *Ursus spelaeus*.

(a)

(b)

(c)

PLATE 10. The exploitation of the pantherins. In the Ancient World big cats alternatively continued to represent power and majesty, and their status also made them the worthy adversary of kings. Their captivity and destruction symbolized the human conceit of power over primal nature. The ancient civilizations of the Near and Middle East revered lions as symbols of power and royalty but, at the same time, relentlessly impacted these and other big cats directly by hunting and indirectly by eliminating their natural habitats. (a) Detail of a life-sized bas-relief of an Assyrian royal lion hunt from the Palace of Nineveh (British Museum, London); (b) Detail of a Roman mosaic depicting a leopard attacked by *venatores* at a stadium; (c) a Chinese Manchu emperor confronting an Amur tiger (painting by Charles R. Knight). *Source:* (a) courtesy of the British Museum, London; (b) courtesy of the Villa Borghese, Rome; (c) courtesy of Rhoda Kalt.

(a)

(b)

PLATE 11. "Tiger farming" in Asia.

To ensure the continued operation of the tiger body-parts trade (and its own profits), the Chinese government's State Forestry Division sanctions, invests in, and promotes the private operation of tiger-breeding facilities, where the animals are often forced to live in filthy enclosures (a). Some tourist operations in other countries, like the now-closed Tiger Temple in Thailand, once secretly sold the body parts of adults and cubs, here seen being confiscated by Thai wildlife officials (b).

PLATE 12. Durga.

Known in Hindu Mythology as Shakti, Devi, Parvati, Ambi, Kali, and by other
names, multiarmed Durga is often shown riding a tiger or lion. She is the fierce
embodiment of the protective mother goddess and turns her wrath on the greed
of those who threaten to destroy natural creation.

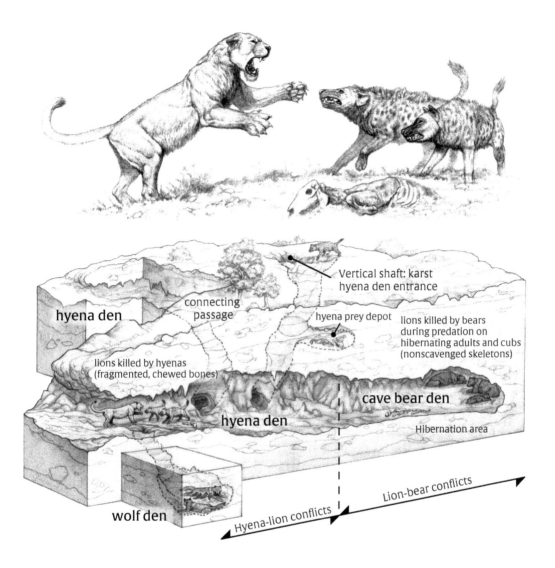

FIGURE 5.9. Composite cave showing conflict zones.

As seen in this diagram based on a composite of excavated sites in the Czech Republic and Romania, European mid- to late Pleistocene caves were periodically occupied by hyenas, cave bears, and sometimes wolves and dholes. While hyenas used areas near the entrance year-round as dens and prey depots, cave bears sought the further recesses to use only during the winter as hibernation chambers and maternity dens for protection from other carnivorans, notably steppe lions. These, and sometimes leopards, periodically invaded the cave to kill hibernating cave bear adults and cubs, coming into conflict not only with the bears but also the resident hyenas, from whom they might try to steal carcasses but for whom they sometimes ended up as prey. *Source:* Modified from original graphic concepts by Deidrich (2014).

the deepest part of the cave. The skeleton's associated and intact state is explained by the fact that cave bears, as committed herbivores, would not have scavenged the carcass. Hyenas, which certainly would have scavenged the corpse had they been able to access it, may have been unable to climb or leap to reach it or were deterred by the closeness of the living bears. The partially disarticulated intact lion skeletons, on the other hand, are the result of the occupational habits of the cave bears, which in their comings and goings through the years gradually but unintentionally moved the loose bones out of their paths into side pockets and recesses with their huge, shaggy paws.

Fragmented, incomplete lion skeletons in caves tell a different story. Eurasian "cave" hyenas, (*Crocuta crocuta spelaea*) habitually denned in caves but, in contrast to the bears, did so year-round instead of on a seasonal basis. At some European caves fossil bone evidence shows that hyenas often habitually favored certain subrecessed areas as maternity dens (indicated by adult female and cub skeletal parts), "commuting" dens (with adult and subadult remains), and "prey-depot/feeding" dens (where bones of prey species and solitary hyenas are found). The tendency of hyenas to establish these areas close to the entrances of caves, rather than in the more remote areas, minimized conflicts with cave bears but not with lions. In Zoolithen-höle and other caves, the bones of steppe lions near the entrance areas show the distinctive chewing, cracking, and crushing damage of hyena teeth; lower jaws were commonly disarticulated from the head by biting through the mandibular ramus to access the tongue. These bones represent both female and male lions and are from individuals that were killed while trying to steal from the hyenas' prey-depot/feeding dens, thereby ending up as food. Partially healed, bite-trauma damage to the skulls of both lions and hyenas from these locations is found on or near the sagittal crest, the likely result of a previous fight with another lion or hyena.

Outside the caves, both *Panthera fossilis* and *P. spelaea* would have had to constantly compete with hyenas, trading the roles of active predator and scavenger depending on the situation—as is the case with lions and spotted hyenas on the Serengeti today. Evidence of these conflicts is provided by excavations at open-air Weichselian glacial and some Eemian interglacial localities, notably in Germany (figure 5.10). One, the Neumark-Nord Lake 1 site east of Leipzig was, during most of its depositional history, an interglacial lake surrounded by a mixed forest of oak, hazel, and yew and attracted megafauna such as red deer (*Cervus elaphus*), Merck's rhino (*Stephanorhinus kirchbergensis*), aurochs cattle (*Bos primigenius*) and straight-tusked or forest elephant (*Palaeoloxodon antiquus*). Periodic algal blooms sometimes, however, made the water fatally toxic for large mammals, and on at least one occasion several forest elephants died in the shallows, probably after drinking. Shortly afterwards, hyenas and steppe lions moved in to compete over at least two partially submerged carcasses. In addition

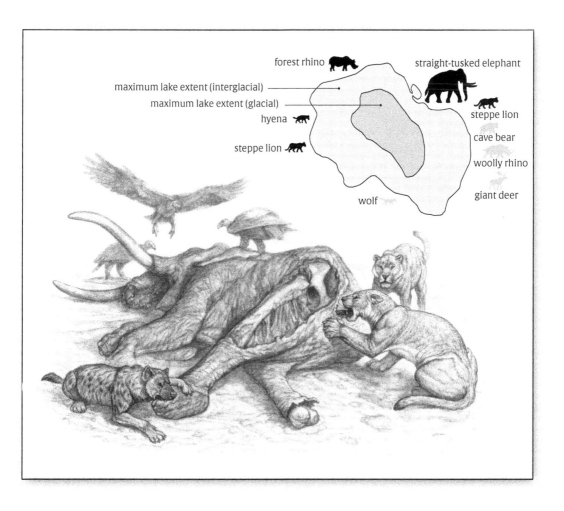

FIGURE 5.10. Neumark-Nord fossil locality, Germany.
The Middle to Late Pleistocene lake in the Geisel Valley of Saxony, Germany, preserved animal and human activities during a 290,000-year interval from the Holsteinian interglacial through the early Weichselian glacial. The lake was more extensive during the interglacial episodes than in glacial ones, and fossil finds show that it was visited by mammals characteristic of each period (gray figures are interglacial, black figures, glacial). In this scene a straight-tusked elephant (*Palaeoloxodon*) has died on the lakeshore during the Eemian interglacial and is being scavenged by hyenas, steppe lions, and vultures.

to the traces found on the fossil bones, insight into these ancient predators' interactions are gained from **taphonomy** field studies of modern African elephants, lions, and hyenas. Although their strong teeth and jaws enable them to start feeding from anywhere on an elephant carcass, modern hyenas often begin by completely eating the feet, whereas the lions with less powerful jaws will choose a soft location, the anus, to access the body cavity with its easily consumed viscera like the heart, liver, and kidneys. In addition to missing foot bones, deep bite and scratch marks on the ventral surfaces of the fossil elephants' lumbar vertebrae and pelvis suggest that one or both predators targeted the musculature of these body areas at some point after the organs were consumed, possibly because they were still soft and pliant (especially important in the case of the lions' more limited chewing abilities). This may have been the point at which lion-hyena conflicts escalated, and during one encounter a female lost her life to a hyena clan during a fight over the carcass. Found among those of the forest elephants, her skeleton speaks of a hard life, even for a lion (figure 5.11). A middle-aged female whose cranial bone fusion and tooth

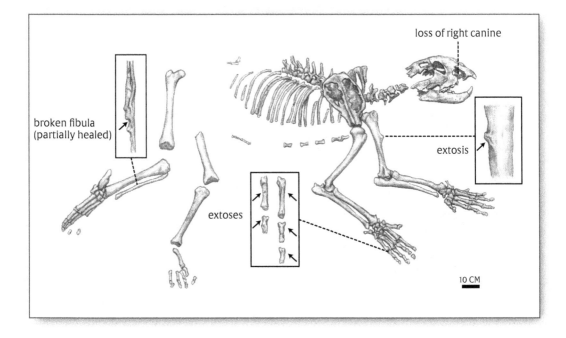

FIGURE 5.11. Neumark-Nord lion skeleton.
Among the lion bones excavated at the Neumark-Nord Lake locality was the articulated skeleton of an adult female *Panthera spelaea*, found among those of straight-tusked elephants, *Palaeoloxodon*. She had suffered serious injuries during her lifetime, including the loss of a right upper canine (because of fracture and abscess, possibly from a prey animal's kick), a broken right fibula, and extoses or outgrowths of the left humerus and right metacarpals associated with arthritis. The diseased and weakened female may have been one of the losers during conflicts between steppe lions and hyenas over carcasses. *Source:* Skeleton adapted from Deidrich (2011a).

wear show that she was between five and eight years old, suffered from arthritic **extoses** or pathological bone growths on her right humerus and some of her left metacarpals, as well as a partially healed fracture of her left fibula, possibly from a kick from a horse or a hyena bite. An inflamed alveolar area around the right upper canine from another undetermined injury would have placed her in constant pain. The Neumark-Nord lion was, if not totally unable to hunt with her pride, clearly limited in her abilities and was forced into confrontational scavenging with hyenas and other predators. In spite of this her situation was not unique, and three other skeletons recovered from other open air sites, one in association with the woolly mammoth *Mammuthus primigenius*, suggest that hard times in securing enough ungulate prey could push steppe lions into deadly encounters with hyenas.

Apart from predator-predator conflicts, caves could be fatal to Pleistocene cats in other ways as well, and as with deposits of this age in Eurasian and African caves, those of South and North American caves help to fill in the unfortunately patchy records of felids on these two continents. In the southwestern Appalachian Mountains near Sweetwater, Tennessee, spelunkers in 1939 discovered the articulated, partial skeleton of the large Late Pleistocene North American jaguar, *Panthera onca augusta*, in Craighead Cavern, an 800-foot-deep, semisubmerged cave. More bones of the individual were found the next year, along with those of its prey, a Virginia deer (*Odocoileus virginianus*) fawn, as well as actual jaguar footprints in the soft cave floor sediments (figure 5.12), whose sizes matched that of the skeleton. These record the last actions of the cat in life: becoming lost and disoriented in the gallery where it brought its prey, it died of thirst and hunger there without rediscovering the entrance to the outside. This may have been a common occurrence for individual big cats in unfamiliar caves, and a similar fate befell a South American jaguar in recent times. The explorer and zoologist Ivan T. Sanderson investigated a cave opening he discovered in a South American rain forest during the last century, worming his way until the passageway finally opened into a large cavern. He fumbled with his flashlight for some moments, but when it came on a wave of fear swept through him. On the far side of the cavern reclined an adult jaguar, its head resting on its forepaws, seemingly awake. After some time he realized that it was dead and, on hesitantly crawling forward, found it to be a mummy, preserved perfectly in the cool, dry air. As with the *P. o. augusta* in Tennessee thousands of years earlier, the cat had apparently tried but failed to find an exit after it entered and, as Sanderson noted afterward, "resigning itself to its fate, it had quietly and serenely composed itself for death."

Natural Trap Cave in the Bighorn Mountains northeast of Lovell, Wyoming, is a 26 m (85 ft.) sinkhole cave that preserved a rich diversity of Late Pleistocene and Holocene mammals, ranging from packrats to

FIGURE 5.12. Jaguar footprints in a cave.
A few sheltered caves in the midwestern United States and in central Europe have beautifully preserved footprints of large ice-age pantherins like jaguars and steppe lions from when they padded across the soft sediments. *Source:* Photo by Bob Osburn.

mammoths, over the millennia, and the talus slope underneath the hole preserved not only a fully complete, articulated skeleton of an American lion (*Panthera atrox*) but also bones representing several individuals of the American cheetah (*Miracinonyx trumani*). This is the only site in which the American cheetah is known from more than one specimen. The American lion specimens from Natural Trap provided a valuable assemblage for comparison with that of Rancho la Brea and are featured in a 2009 paper by H. Todd Wheeler and George T. Jefferson that assessed the proportions, sexual dimorphism, and behavior of *P. atrox*.

During the Late Pleistocene, the huge American lion successfully penetrated as far south as Chilean Patagonia in South America. In northern South America it was an apex predator, sharing this role with the jaguar *Panthera onca*, the puma *Puma concolor*, and the massive dirk-toothed sabertooth *Smilodon populator*. At one particular locality, the Cueva del Milodón, near Última Esperanza in Chilean Patagonia, northwest of Puerto Natales, the preservation of some lion remains, as described in a 2017 paper by the researchers Nicolas R. Chimento and Federico L. Agnolin (both Museo Argentino de Ciencias), offer a unique glimpse of how this species may have looked and possibly even behaved in life. The remains (originally reported by Lehmann-Neitsche in 1899 and incorrectly assigned to an extinct jaguar, "*Panthera onca mesembrina*") include pieces of fur-covered, mummified skin "adhering to the 'face' (maxilla?) and forelimb," as well as an isolated phalanx with preserved keratinous claw. The color of the skin patches, as reported in a paper by Roth (1904) is "a reddish brown tone (rufous), a color that constituted the background of available skin patches from the limbs and body." Although it isn't clear whether Roth refers to the color of the fur or the skin, the coat hairs as preserved are in actuality a light tan color. Roth further speculates that this "indicates that the species probably exhibited dark and some yellowish color stripes, at least in the forelimbs." The same cave preserved several skulls of the ground sloth *Mylodon* with deep tooth gouges in the posterior parietal region. These indicate an attempt to rend away the temporalis muscles as a food source, most likely by a large felid like *P. atrox*. There was also a coprolite (fossilized fecal mass) that contained a large number of *Mylodon* dermal ossicles.

In 1997 L. A. Borrero reported another fossil site near Cueva del Milodón, which he described as a deep cave and "burrow" that contained a bone accumulation of large mammal species, including *Mylodon*, the "dwarf horse" *Hippidion*, the huemul or South Andean deer *Hippocamelus*, the camelid *Llama*, and others. Many of the bones were broken and exhibited large felid-sized tooth marks. The transport of prey to rocky areas or caves is typical of pumas or leopards but not of extant lions, and, if actually representing a typical cache or accumulation of *P. atrox* prey, this would imply a different predation behavior for this species as opposed to extant lions, which invariably consume their prey in the open.

The cave of Gran Dolina in the Sierra de Atapuerca region near Burgos, Spain, is Europe's oldest continuously human-occupied site and the second-oldest European habitation after Dmanisi, Georgia. Bands of *Homo antecessor* or even the earlier *H. erectus* began living there about 780 ka, leaving charcoal hearths, tools, and their own bones, some showing the unmistakable cutting marks of defleshing and cannibalism; its occupation ended with *H. heidelbergensis* about 300 ka. That these ancient Europeans were skilled hunters who captured a variety of ungulate prey is known from the butchered bones littering the cave floor, but the condition of those belonging to the early steppe lion *Panthera fossilis* was completely unexpected. The limb bone joints have artifactual cut marks indicating butchering and defleshing like that of the other prey animals, and some of the long bones show deliberate cracking and splitting to get to the marrow. That the big cat was actually eaten by the cave's human inhabitants is beyond question, but why? *P. fossilis* was a formidable adversary to confront and kill, and the strong, gamey taste of most carnivoran flesh would make this meat much less attractive than that of an herbivore. One answer may be that it simply represents a chance encounter between the big cat and human hunters, with the latter simply being unwilling to let the meat go to waste. On the other hand, for early Europeans steppe lions might have had a symbolic importance. Perhaps, as in modern ethnic societies such as the Masai of East Africa, to kill such a ferocious animal conferred status, initiating a young male as a formally recognized adult and tribal member. A lion hunt might have been followed by bringing the corpse back for ritual consumption, allowing the hunters to acquire the lion's attributes.

Although steppe lion hunting by Paleolithic humans may not have been a regular occurrence, the intriguing find of at least nine bones of *Panthera spelaea* on the floor of the Middle Magdalenian (14.5–12 ka) site of La Garma cave in the region of Cantabria, northern Spain, points to a direct use other than for food. These are all from the distal phalanx (the one that bore a claw in life) of either the manus, pes, or both, and each has repetitive, diagonal cut marks across the articular surfaces indicating that a blade was used to separate it from the adjoining phalanx. The pattern of cuts is practically identical to those used by modern furriers or taxidermists when skinning a carnivoran's foot and indicates that the intention was clearly to keep the claws attached to the pelt. That the phalanges were found arranged in a roughly circular pattern on the cave floor has suggested to some archaeologists that the claws were a part of a pelt that functioned as a "rug" within the cave and were the only objects left when the pelt eventually degraded (figure 5.13). If so, such a rug may have had a ritual meaning for those who sat or lay on it. Bones with artifactual cuts found at other Magdalenian cave sites in the Swabian Jura district of southern Germany suggest that there, too, steppe lion hunting was a part of late Paleolithic culture. The fact that modern cat fur "slips" or easily comes away from its pelt after time and has inferior insulation qualities compared to that of other carnivorans like bears or wolverines, means that a steppe lion pelt rug, if similar to modern cat skins, was not used for practicality or warmth. Could it have had instead a symbolic or honorific meaning?

On the chilly evening of December 24, 1996, the first human voice heard in almost 30,000 years echoed within an obscure cliff-side cave in the Gorges de l'Ardèche, southeastern France. "They have been here!" shouted Eliette Brunel to her other two spelunker friends as she first glimpsed the drawings on the cave wall. These were only the first of a spectacular menagerie of some 400 late-ice-age animal paintings, rendered variously in charcoal and red ocher, that went on from one cavern to another, ending with perhaps the most dramatic of all, named the Gallery of Lions (plate 6b). Appearing to follow a herd of bison, a pride of sixteen *Panthera spelaea* steppe lions had been portrayed in charcoal lines with such grace and power that it literally brought the onlookers to tears when they

FIGURE 5.13. Steppe lion phalanges from La Garma cave.
The discovery of the terminal phalanges of adult steppe lions in caves such as La Garma, Spain, indicates that these had a special significance for these Middle Magdalenian hunters. Although the claws were not preserved, the cut marks at the base of the phalanges suggest that they were intentionally kept as part of preserved paws or were from a skin that may have been used as a ceremonial rug.

first saw it. Until its discovery the caverns, whose entrance was blocked by a massive rockslide that occurred about 29 ka, was occupied at different times by cave bears and by Aurignacian-culture humans, who left behind their horsehair paint brushes, hearths, and sometimes finger prints on the walls. Lions appear frequently in the 36,000-year-old art of Chauvet-Pont-d'Arc Cave, and this has suggested to some that these animals may have been particularly special to the humans that depicted them. Their distinctive profiles could mean that the artists who created their likenesses knew them as individuals, much in the same way that the naturalist George Schaller came to know territorial pride members of modern African lions in his field studies. From the drawings it seems that the artists were also careful observers regarding anatomy. The males, unlike modern African lions, were almost or totally maneless, with their sex clearly indicated in other caves by similar figures with scrotums. In the savanna lion, *Panthera leo*, a dark, luxuriant, and proportionally large mane is correlated with sexual potency in the form of higher testosterone levels and, by its size, a way of intimidating other males. If steppe lion males as well as females had actively hunted in snow-covered terrain, however, a mane would probably have been a detriment by making them highly visible to prey. The cave art also suggests that like adult savanna lions, steppe lions apparently had no obvious coat patterning. Although *P. spelaea* may have had a monochromatic coat, as winter approached this may have changed to a much lighter, paler color, which along with the lack of a dark mane would have given it an advantage in avoiding being spotted by prey.

The Aurignacians of Chauvet-Pont-d'Arc might have felt a connection with steppe lions that went beyond simple admiration. At least one drawing seems to bear the head of a lion with the body of a male human, an image known as a therianthrope, or "beast-human," uncannily anticipating by thousands of years Sekhmet, the lion goddess of ancient Egypt who presided over fire, war, fertility, and healing. In the cave of Hohlenstein-Stadel, Germany, a small mammoth-ivory carved figure of a tall human man with the unmistakable head of a lion was found in 1939 (plate 7d), while other originally full-figure carvings of robust big cats, some of which have carefully etched cross hatchings and pits, come from Vogelherd Cave, both in the Swabian Alps (plate 7a–c). Similar figures have been found in French caves. Although perhaps not regarded as "gods" in the ethnological sense, steppe lions certainly must have inspired profound feelings of respect, power, and awe in our early ancestors as they watched the animals stalk, kill, and procreate. Jean Clottes (French Ministry of Culture), a specialist in archaeology and rock art, believes from his research studies that Paleolithic humans lived simultaneously in two worlds, the first being one of the senses and the second a spiritual dimension that lay beyond everyday consciousness. As in cultures like the Mayan, caves were portals to inner realms where the spirits dwelled; perhaps murals like those of Chauvet-Pont-d'Arc had the ability to give these invisible entities a form that allowed humans to commune and, possibly aided by hallucinogens, receive guidance from those beings on which they depended for their very lives, the animals. Few moments in nature are as dramatic as the act of predation, and steppe lions, with their stealth, grace, and ability to hunt cooperatively and take down large and occasionally even huge prey, may have been at the pinnacle of animals to honor and emulate. By incorporating the images of carnivoran and human into a single artifact, the Aurignacians may have believed they could transcend their human limitations to achieve oneness with the lions and acquire their skill, power, and hunting success (figure 5.14).

FIGURE 5.14. Aurignacian human carving lion figure.
Perhaps intended as spiritual icons or simply created out of sheer admiration, small effigies of steppe lions, leopards, and other animals, sculptures known as portable art, were created by Aurigacian and Magdalenian hunters of France and Germany.

To view the steppe lions and other animals of Chauvet-Pont-d'Arc is to be struck with a sense that one is in the presence of something profound, of incredible visions that came from the union of ancient human minds with a natural world of animals and ecosystems that are no more. At the edges of this sensation lingers a subtle sense of regret. The magnificent creatures that our distant ancestors must have seen every day were to be largely gone in a few thousand years. How did this happen, and what kind of world did the surviving felines inherit?

CHAPTER 6
Aftermath of an Ice Age

Eastern Beringia, Canada, some 11,500 years ago: The spear has found its mark, and the reindeer is dead. Hefting the hide, meat, and organs on their shoulders after butchering the kill, the hunters—a weathered man and his eight-year-old son—are well on their way home when the father remembers the den. Located under a willow-sheltered embankment on a river terrace, it is the place where, several months earlier, he spotted a female lion bringing in mouthfuls of grasses and moss after digging a space in the gravelly soil to prepare for the birth of her cubs. The older hunter has noticed few prides or even individual steppe lions over the last year and, wanting his son to finally see one, he silently motions for the boy to carefully crawl to a downwind knoll. The wind carries the faint sounds of tiny mewling cries. After a time the lion emerges, her pale yellow, brindled fur ruffling in the chilled spring air. After looking warily around, she silently pads down the terrace to locate her pride along the thawing river.

Although they cannot know it, the man and his son will be the last humans to see a living steppe lion (figure 6.1).

. . . . .

On a late August morning in 2015, a group of commercial fossil collectors in Siberia was probing a frozen embankment in the Edoma permafrost deposits several meters deep for mammoth bones and tusks when they came across two small forms. Although they did not immediately recognize what they were, the diggers later realized that in the depression lay a pair of mummified *Panthera spelaea* steppe lion cubs, perfectly preserved by the permafrost after their den had suddenly collapsed on them 25,000 to 55,000 years earlier during the Karginskii interstadial (figure 6.2a). The little bodies were placed in a nearby glacial outcrop to keep them frozen while scientists in the Academy of Sciences of the Republic of Sakha, Yakutsk, were notified. At the lab the babies, which were named "Uyan" and "Dina" after the Uyandina River near their discovery site, began to reveal their secrets. The cubs' eyes were still closed, and their deciduous or milk teeth, almost but still not quite erupted from the gums, placed them at between two and three weeks old at death. Housecat-sized Uyan, at 2.8 kg (6 lb.). would in life have been 2.1 kg (4.6 lb.) heavier than a modern African lion cub of the same age,

FIGURE 6.1. Father and son watching female *Panthera spelaea* lion entering den.

(a)

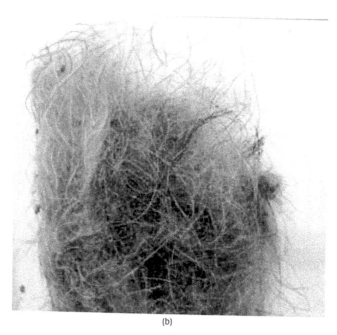

(b)

FIGURE 6.2. *Panthera spelaea* mummified cub.
Buried for millennia in its icy den, a recently recovered mummified *Panthera spelaea*
cub (a) may offer new information about this now extinct pantherin species.
(b) Previously, samples of fur had been found with the bones of adults in caves.
Source: Cub photo (a) by Vera Salnitskaya, courtesy of Yakutian Academy of Sciences;
fur photo (b) courtesy of Barnett et al.

and both had proportionately shorter tails at 7 cm (3 in.). The cubs were covered in dense, 3 cm (1.2 in.) yellowish-tan fur, matching that found in association with adult bones in other sites. A CT scan showed that Uyan's stomach had contained milk a few hours before death. The cubs' skulls were partially crushed and some of the neck vertebrae separated, which probably occurred when a springtime thaw caused the permafrost in the soil of the walls to melt, causing the collapse of their den. To everyone's disappointment the quality of the cubs' ancient DNA (aDNA) was poor, ruling out the hopes of a South Korean scientific team that significant steppe lion genes could be replicated, amplified, and studied. The comparative uncommonness of large apex predators in living ecosystems, let alone in the fossil record, however, makes the cubs an amazing scientific treasure of information. At the time of this writing further investigations are underway that may tell us much more in the future, especially as a new, 50,000-year-old cub named Boris, estimated at one and a half to two months old at the time of death, was discovered on the banks of the Tirekhtykh River in the Abyisky District of Yakutia in Russia's Sakha Republic. In 2018 yet another, Spartak, was found about 15 m (49 ft.) away from Boris.

In addition to the four steppe lion cubs, other Pleistocene mammal corpses have recently begun to emerge from the Russian, Alaskan, and Canadian permafrost as the climate warms. As well as pointing to the mortality of young animals in such a harsh environment, these speak of a lost faunal richness now seen only in East Africa but typical of Holarctic lands until about 14,500 years ago. At that time that the climatic transition from the latest Weichsel (or Würm) glacial into the present Holocene interglacial was well underway, and the northern continents were rapidly losing most of their largest and most distinctive herbivore and carnivoran species (but also some small ones as well). Although megafaunal extinctions had occurred in Australia many thousands of years earlier, Europe, Asia, and Africa now also experienced population decline and then die-offs in some **keystone** mammal species. These had survived the nine earlier, similar transitions from glacials to interglacials, some of which were intense, over hundreds of thousands of years. Whereas many of Europe's most iconic ice-age animals, such as woolly mammoths, steppe bison, woolly rhinos, cave bears, hyenas, and sabertoothed cats were either extinct or almost extinct by an earlier date, the megafauna of the New World remained spectacular until about 12–11 ka (plate 8 c-e). The roster among the North American felids alone is impressive: smilodontin and homotherin sabertooths (both apparently extinct elsewhere), American lions, jaguars, pumas, bobcat, and lynx. Debate still continues over the reasons for the end-Pleistocene extinctions, but recent data from genetic studies and isotopic research, as well as new fossil discoveries, are starting to bring into focus both the ultimate causes and a timetable for the extinctions, especially in the case of felids. Although many pantherins present at the end of the Pleistocene still survive as species, the steppe lion *Panthera spelaea* and American lion *Panthera atrox* are extinct, as are their sabertoothed relatives, and an understanding of the causes lies at the roots of saving modern big cats.

The natural extinction of a species in the geological record is considered by some a sign of a flaw or deficiency in the animal's makeup but is a normal phenomenon—usually the outcome of a complex series of circumstances to which the creature is unable to adapt. When long- or short-term changes in the environment occur, it becomes a race against time to adopt successful new behaviors and evolve compatible physical changes that will allow populations of a species to reproduce fast enough to outpace accelerated mortality. If the species is adaptable it may survive and continue; if not, it may flicker out and die after surviving for thousands or even millions of years. The continued presence of big cats in the modern world, and the absence of their ancient counterparts the sabertooths, is attributable to a set of environmental changes that may also have included prehistoric humans as the "wild card" in the game of survival and extinction. As described in chapter 4, homotherin sabertooths succeeded in becoming successful apex predators in open environments by combining the machairodontine rapid method of dispatching prey with a social pride or clan behavior that allowed them, like steppe lions and hyenas, to take down large prey and defend kills. Based on this, Mauricio Antón, Angel Galobart (Institut de Palaeontologia), and Alan Turner in 2005 presented a convincing scenario of competing social predators during the European Pleistocene, one in which the early human *Homo heidelbergensis* was the decisive factor in the outcome.

Homotherin ("scimitar-toothed") sabertooths such as *Homotherium latidens* readily adapted to predation opportunities presented by large-bodied, open-habitat ungulates at about the same time as steppe and savanna lions, between 2.0 and 1.5 Ma. By around 400 ka, however, these sabertooths had become extinct in Eurasia except for possibly some late surviving populations in western Europe. Like lions, when making a kill they probably did not use all the available meat from a carcass and certainly not the in-bone tissues like marrow, which made scavenging from their kills and those of lions an increasingly attractive source of protein for social scavengers like early hominins and the bone-crushing forms of hyenas. Although the australopithecine hominins probably practiced nonconfrontational scavenging (safely taking meat after the apex predators had left the kill), as opposed to confrontational scavenging (directly challenging them for its possession), this situation gradually changed as later, more advanced forms developed better weapons and cooperative strategies. By the time of *Homo heidelbergensis* (see later in this chapter) all three carnivoran apex predators—lions, homotheres, and hyenas—were likely to be

FIGURE 6.3. Confrontational scavenging between *Homo heidelbergensis* and homothere. Because of their dentitions, homotheres and other sabertoothed cats used less flesh from a carcass than did pantherins, providing early humans like *Homo heidelbergensis* with opportunities to scavenge meat by actively driving the cats away from their kills.

confronted by upright, determined, and well-armed primates who wanted the meat for themselves (figure 6.3).

Anatomical studies by Antón and associates demonstrate that, compared to pantherins, homotheres had shorter backs and, unlike smilodontins, more gracile, less muscular forelimbs. Although heavier forelimbs and short backs gave sabertooths in general an advantage in grappling with and overcoming prey, it also caused them to be less effective than lions and other cats in leaping and rapid acceleration. Unlike other sabercats and pantherins, lighter forelimbs also made homotheres less capable of killing a large animal on their own, and as a result these sabertooths would have hunted larger prey more as a group than individually. In each of their respective clades, both machairodontines and pantherins generally overlapped in size. Then as now, most large predatory species in a given ecosystem probably chose the most abundant suitably sized prey. Whereas *Homotherium latidens* and *Panthera spelaea* must have broadly overlapped in prey size preferences, the homothere's smaller size, lesser weight, and less

FIGURE 6.4. Hyena clan driving homothere away from its kill.
During the Middle Pleistocene Eurasia had at least three and possibly four kinds of social carnivores that competed for prey: humans, steppe lions, hyenas, and perhaps homothere sabertooths. Although formidable hunters, homotheres were not as physically powerful as lions and, if outnumbered as a group, would have relinquished their prey to giant hyenas.

powerful forelimbs and bite would have placed it at a disadvantage had it been surrounded by lions. And with an estimated adult weight of 127 kg (280 lb.), it would likewise have been the loser in a face-off with the similarly sized but more powerfully built giant short faced hyena *Pachycrocuta brevirostris* with an estimated weight of 113 kg (250 lb.). When giant hyenas became extinct and were replaced by their smaller relative, the modern spotted hyena *Crocuta crocuta*, homotheres gained a size and weight advantage and confrontations over prey came down more to a question of numbers (figure 6.4). Homothere hunting techniques worked best in open environments but, when faced with more and more competition from lions and hyenas during the mid- to late Pleistocene, the homotheres may have begun to habitually seek some undergrowth or other cover from which to make their kills, becoming increasingly marginalized from their former role as dominant predators.

Previously, the Eurasian smilodontin sabertooths like *Megantereon* (plate 8a) had become extinct, perhaps from competition with jaguars, tigers, and leopards, and by about 400 ka first *Homotherium hadarensis* in Africa, then *H. latidens* in Europe (plate 8b), and *H. ultimum* in Asia came under serious competitive pressure. It was about this time that *Homo heidelbergensis* (figure 6.5) dispersed from Africa to make its appearance in

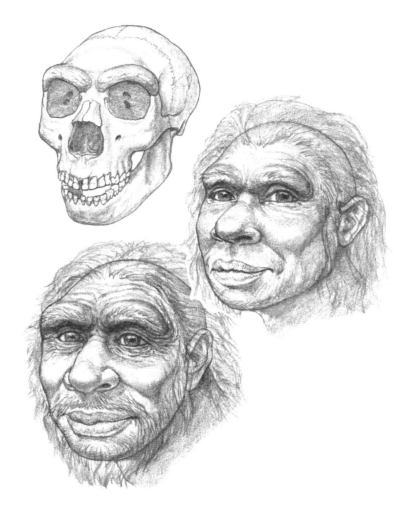

FIGURE 6.5. Skull, life restoration of male and female *Homo heidelbergensis*
Known from beautifully preserved skulls and other remains at the Sima de los
Huesos Cave in Spain, Petrolona, Greece and other localities, *Homo heidelbergensis*
is characterized in part by its massive brow ridges and very wide nasal aperature.
The first known cranium was from the Kabwe Site in Zimbabwe ("Rhodesian Man"),
and is sometimes considered to be a representative of archaic Neanderthals.
The skin color, hair texture and other details of this restoration are conjectural,
and the hair has only been lightly suggested so as to reveal the contour of the heads.

Europe. This hominin was no longer an opportunistic, confrontational scavenger but a true hunter of ungulate prey, well advanced in blade-making and spear-fashioning capabilities, and it now became a part of the Eurasian predatory guild, increasing the pressure on the existing predators as another hunter in the carnivore guild competing for the same kind of prey. The continuance of the Eurasian steppe lions, hyenas, and hominins but the disappearance of further homotheres from the fossil record indicates which species lost. After some 2.0–2.59 million years, these unique, highly specialized sabertooths became extinct in the Old World, at about 28 ka, except for some relict populations that may have survived until sometime in the Later Paleolithic. It appears that the entrance of modern humans into the ecosystem pushed *Homotherium* over the brink.

With the extinction of the homotheres, pantherins were left as the supreme felid predators of Eurasia. In addition to the sabertooth extinctions, steppe lions by this time were also rid of another competitor, the giant hyenas *Pachycrocuta brevirostris* (in Europe) and *P. sinensis* (in Asia), though they still had to contend with the smaller and more nimble "cave hyena" *Crocuta crocuta spelaea.* In Eurasia steppe lions also had to compete with modern wolves (*Canis lupus mosbachensis*), which had been a part of the predatory guild since middle Pleistocene times and whose prey preferences probably overlapped those of steppe lions and spotted hyenas in taking down horse, reindeer, and bison. In spite of this competition, both steppe and savanna lions, along with tigers, jaguars, leopards, short-faced bears, and polar bears, shared the role of the late Pleistocene's top terrestrial predators. Steppe lions and savanna lions occupied a huge geographic range, extending from northern Britain to southern Africa and eastward from Spain to the Canadian Yukon. Steppe lions were versatile hunters, both individually and in groups. From the mid-Pleistocene onward they uneasily but successfully shared the vast, open plains with hominins and other social predators, intercepting the migrating herds that came through their territories. Steppes and savannas both now and in the past support a greater biomass than any other ecosystem, and the big cats overlooked what must have been rich hunting grounds. Why then did the steppe lion become extinct?

The answer to the mystery of the steppe lion's extinction could be rooted in one—or possibly all—of four hypotheses. The first is the contraction at the beginning of the Holocene of the mammoth steppe and other habitats to which the steppe lion and its preferred prey were adapted. The second is that the cat died out because it was heavily reliant on a particular prey species that became extinct, temporarily rare, or regionally extinct from environmental stress caused by the climatic change. The third is a decrease in genetic diversity that rendered the species vulnerable. The fourth is its replacement in the carnivoran guild by humans. Let's examine all these possibilities in turn.

During the last glacial maximum, the mammoth steppe was the earth's most extensive biome. It stretched from Spain eastward to Canada and from China northward to the Arctic Islands. Its vegetation was dominated by high-productivity grasses, herbs, and willow shrubs, and the animal biomass was dominated by bison, horse, and woolly mammoth. The mammoth steppe had a cold, dry climate. During previous interglacials, trees and shrubs expanded northward but left northern Siberia and Beringia (Alaska and the Yukon) as a refugium of mammoth steppe vegetation. At the beginning of the Holocene the climate switched to a warmer, wetter one, and higher temperature and rainfall accelerated the invasion of the dry grasslands by mossy forests, tundra, and muskeg or wetlands. The steppe bison (*Bison priscus*, plate 9a), which survived across the northern region of central eastern Siberia until 8,000 years ago, exemplifies the effect of the warming on the mammoth steppe biota. Stomach contents of a frozen bison mummy from northern Yakutia (Russia) indicated that it was a pasture grazer in a habitat that was becoming dominated by tundra and shrub vegetation. Continuing decrease of its preferred habitat led to population fragmentation and ultimately to extinction, and this would have had profound consequences for a predator that might have been primarily dependent on bison, secondarily on horse (*Equus ferus*), reindeer (*Rangifer tarandus*), and occasionally cave bear (*Ursus spelaeus*) as prey sources (plate 9b, c, d).

A study of the spacing of tooth marks on the hide of an Alaskan mummified steppe bison, "Blue Babe" (figure 6.6), by Dale Guthrie (University of Alaska) confirms that Beringian steppe lions definitely preyed on bison, a conclusion confirmed by the isotopic analyses of Kirillova and associates (see chapter 5). If bison were indeed a major source of prey for *P. spelaea*, their regional or local extinction would have meant serious trouble for the cats, which would have been forced to focus on other and less abundant game. Although the steppe lion was probably as versatile as the savanna lion in its hunting ability, both individually and as a group, the prey spectrum available to lions in East Africa, with its considerable numbers of ungulate species, is greater in diversity than that of the known species of the Eurasian mammoth steppe. This broad prey spectrum acts as a buffer for East African lion populations in areas like the Serengeti where, depending on the time of year and on opportunistic encounters, they preferentially feed on species like Burchell's zebra (*Equus quagga burchelli*), wildebeest (*Connochaetes taurinus mearnsi*), and Cape buffalo (*Syncerus caffer*) but also have access to about fourteen other ungulates. Smaller species offer savanna lions other options. The warthog (*Phacochoerus africanus*) has a high reproductive potential and is a staple of resident savanna lion populations during the dry seasons when migratory grazers are absent. While the warthog's northern counterpart the Eurasian wild pig (*Sus scrofa*) could have been expected to provide a similar alternative prey base

FIGURE 6.6. Mummified steppe bison (*Bison priscus*).
Nicknamed "Blue Babe" after Paul Bunyan's legendary ox, an excellently
preserved carcass of the steppe bison (*Bison priscus*), described by Dale Guthrie
and now an exhibit at the University of Alaska Museum of the North, Anchorage,
was found in permafrost in 1979. Bearing toothmarks that match the size and
spacing of Beringian steppe lions, this provides a rarely encountered example of
direct predation by an Ice Age pantherin. *Source*: Photo by Nigel Collingwood,
courtesy of the University of Alaska Museum of the North.

for steppe lions, the fact that these pigs are forest and not open-country
dwellers decreased their availability. The problem of limited prey options
is exemplified by the Canadian lynx (*Lynx lynx*). During the winter this cat
is extremely, almost exclusively (up to 97 percent), dependent on Arctic or
snowshoe hare (*Lepus arcticus*) as a prey source. When the hare's populations soar during productive spring and summer seasons so do those of the
lynx, but when the hare's crash during a severe winter the lynx mortality
from starvation is high.

Besides having to adjust to a diminishing bison prey source, the loss
of open habitats in which they hunted may have also affected steppe
lion survival in the same way that habitat loss affected homotheres in the
mid-Pleistocene. The steppe lion, like the modern Asiatic lion *Panthera leo
persica*, might have been flexible enough to adapt to hunting in open forest

but there would have encountered serious competition from the established apex predators—tigers and leopards. These had benefitted from the Holocene expansion of the boreal forest, a habitat that sheltered typically woodland prey species like deer, woodland bison, elk, moose, and wild pig and in which they were better adapted to hunt than lions. The question then arises: if steppe bison eventually became unavailable to lions through extinction and woodland ungulates were not an option, why didn't other large, plains-dwelling prey species serve as prey substitutes? Several of these (horse, camel, reindeer, yak, and others) continued to survive in large numbers, and we must ask why steppe lions failed to survive by adopting these as prey.

In steppe environments during glacial periods, seasonality and rainfall were critical factors in the survival of these and other herbivores because they affected forage availability. Like their African savanna equivalents, large and medium-sized mammoth steppe ungulates would have partitioned food resources, with each species foraging on a preferred growth stage and migrating to new areas when that resource was exhausted. Modern populations of ungulates, however, tend to be limited less by actual food abundance and more by food quality, and, in the north, winters and dry seasons are especially stressful. Plants become more fibrous and harder to break down within the foregut-dominant digestive systems of bison and other ruminants, the plant digestibility being reduced by 40 to 80 percent. Calorie levels of grass and browse are often low, with most nutrients being stored in root systems and tubers below ground level. In contrast, the hindgut digesters, perissodactyls like horse and rhino, are better able to extract nutrition from dry, calorie-poor grasses. A 2017 study by Timothy Rabanus-Wallace (Australian Centre for Ancient DNA) and colleagues investigated prolonged warming events after the last glacial maximum that led to deglaciation. They documented an increase in rainfall that reached a maximum just before the transformation of rangeland that had supported megaherbivores into widespread wetland that supported herbivore-resistant plants. Over most of the planet, moisture-driven environmental change led to megafaunal extinctions. In Africa, however, the continent's transequatorial (geographically vertical) orientation allowed rangeland to persist between deserts and equatorial forests, so fewer savanna-dwelling megaherbivores became extinct.

In Eurasia and North America, the transition from glacial periods into interglacials (or stadials into interstadials) had a major effect on the mammoth steppe. The interiors of large continents such as Eurasia and North America shifted from an overall "maritime" or "oceanic" climatic system, with a relatively narrow temperature range and consistent precipitation, to that of a "continental" one, with more extreme seasons. Longstanding ecosystems with a mosaic of habitats were profoundly changed by shifts in temperature and rainfall patterns. This in turn had an effect on the dispersals, abundance, and seasonal nutritional value of plants and had a major impact on the distribution and abundance of mammals that fed upon them. In 1988 Hartmut Heinrich (Federal Maritime and Hydrographic Agency, Hamburg) described a previously unknown phenomenon (now known as a "Heinrich event", named after its discoverer) that was demonstrated by climatologists to have occurred several times from the Middle through Late Pleistocene. Heinrich noted that in some Pleistocene sequences of fine-grained deep-sea sediments there were several episodes of unexpected accumulations of pebbles and cobblestones that were much too large to have been transported by ocean currents. These stones had instead been picked up by continental glaciers as they slowly moved across the land toward the coastal edges and were subsequently dropped from icebergs that "calved" or broke off from the glaciers and melted in the open sea. Thus Heinrich events signaled the ends of stadial cold periods and the beginning of interstadial warm phases. This is relevant because normally harsh winter foraging conditions might have become far worse during a glacial-interglacial or stadial-interstadial transition, which, according to some climatic models, would have produced more irregular, high-velocity winds and lowered temperatures. Current evidence suggests that the last event accompanying the beginning of the Holocene (Heinrich 5) was unusually severe and that large ungulate grazers that had succeeded in withstanding previous transitions were unable to survive. When the climate and temperatures altered regional and local plant associations, some formerly sympatric species with overlapping temperature tolerances may have retreated to grow in different, separate areas. This in turn had an effect on the everyday lives of the herbivores that depended on these associations. For a foraging horse, bison, or reindeer, a yearly migration in search of quality grazing or browse that lasted several more days than the previous one and each winter's foraging's yielding less nutrition than the last could have meant the difference between life and death.

Winter at high latitudes usually means low temperatures and snow. Hoarfrost, although normally lasting only one or two days, often coats vegetation with icy sheaths that require a great amount of nibbling to free the grass inside, making feeding more time-consuming. The descendants of the Tibetan Plateau faunas—woolly rhino, horse, bison, camel, yak, chiru, and others—were already well able to withstand extreme low temperatures, having evolved various adaptations such as fleecy undercoats and subcutaneous fat. For modern Central Asian ungulates, however, much more critical than cold is snow-cover depth. Deep snowfall slows and renders movement strenuous, especially when foraging or escaping from predators. Under poor foraging conditions, the expenditure of critical calories each day cancels out the ones derived from food and squanders critical fat

FIGURE 6.7. Mongolian gazelle (*Procapra picticaudata*) floundering in snow.

reserves. After a heavy snowfall, the chances of surviving may depend on leg length versus snow height: 88 cm (35 in.) for a lanky elk (*Alces alces*), 60–71 cm (24–28 in.) for a stockier Prezwalski's horse, and 40 cm (16 in.) for the smaller saiga and chiru. The weight the snow crust can bear without an animal breaking through and floundering is also critical. The broad, splayed hooves of a reindeer exert pressure of only about 0.9 kg (2 lb.) each; the narrower ones of a hemione or wild ass (*Equus hemionus*) exert about 6 kg (14 lb.) per hoof; and those of a Tibetan gazelle (*Procapra picticaudata*), 2 kg (6 lb.): these pressures are often too great for the crust to support (figure 6.7).

A prolonged hoarfrost cover can create a catastrophic event known in historic times as *djout* in Russian or, in Mongolian, *dzud*. This translates as literally "the death of a great many both wild and domestic ungulates," and in severe winters during the late 1950s and early 1960s *dzud*s caused the catastrophic die-off of thousands, sometimes millions of animals from starvation in central Asia. Exceptions were nearby woodland-dwelling species like moose or elk that survived because of their long legs and preferred habitats that provided frost-free bark and conifer browse. If *dzud*s are as infrequent as every eight to ten years then populations can rebuild, but if they recur in two consecutive winters, animals can become locally extinct. As described by Pierre Pfeffer in 1968, following one *dzud* the Altai argali (*Ovis ammon*) died out in Transbaykalia and Ustyurt; the Mongolian goitered gazelle (*Gazella yarkandensis*) disappeared from the Minusinsk steppes on the upper Yenisey River in Siberia; and hemione and wild ass and Przewalski's horse became locally extinct in Kazakhstan. If such local extinctions had struck mammoth steppe ungulate species over consecutive winters during the severe Holocene transition, even temporarily, then *P. spelaea*, which was probably territorial like modern lions (*P. leo*), could well have become extinct.

Felids and other carnivorans may, like herbivores, have been under more intensive survival pressure during some glacial periods, but it was of a different kind. Blaire Van Valkenburgh (University of California Los Angeles) and her colleague Fritz Hertel (California State University, Northridge) compared the frequency of tooth fracture in extant and Pleistocene carnivorans and demonstrated that that the frequency of tooth breakages was three times higher in four extinct species—American lion (*Panthera atrox*), dirk-toothed cat (*Smilodon fatalis*), dire wolf (*Canis dirus*), and coyote (*Canis latrans*)—than in their nearest living ecological equivalents. From this they concluded that the Rancho La Brea species were under pressure to use carcasses more fully than their modern equivalents and likely competed more intensively with each other and other forms such as the short-faced bear (*Arctodus simus*). In Eurasia during this time the equivalents of the short-faced bear and dire wolf were not present, but their places were taken to some degree by the Eurasian brown bear and spotted hyena, the latter a major competitor for carcasses with the Eurasian steppe lion (see chapters 3 and 5). Additional evidence for greater carcass utilization comes from the previously mentioned steppe bison mummy, "Blue Babe," which had a canine tip from a Beringian steppe lion broken off in its hide. This breakage is likely to have occurred during attempts by this individual (or a group) of steppe lions to scavenge after the carcass had frozen, implying that the lions were under pressure to consume the flesh regardless of its hardness. This and the Rancholabrean study imply that life was probably hard for both steppe lions and their prey during intervals of extreme cold.

Although perhaps not directly contributing to steppe lion extinction, a temporary decrease in genetic diversity might have been a second, contributing factor. Based on aDNA samplings from preserved frozen tissue, Ross Barnett and his associates in 2009 were able to identify a major decline in genetic variation, a "bottleneck," in European and western Beringian *Panthera spelaea* populations during an interval from 48 ka to 32 ka, followed by a population expansion. The biological cause of this is presently unknown, but a similar and earlier bottleneck is documented in the steppe bison (*Bison priscus*) between 50 ka and 48 ka and also in woolly mammoths and rhinos. In each case the animals were left with reduced populations and smaller, more fragmented, and less varied gene pools. As with steppe bison, so with steppe lions—these animals needed time for their genomes to once again diversify, and under such circumstances individual populations would be more vulnerable to sudden climatic changes. The distribution of steppe lion remains in Europe suggested to Anthony J. Stuart (Durham University) and Adrian M. Lister (British Museum of Natural History) that there was a "Lion Gap," in which these cats were actually absent from Europe for the 11,000 years between 29 ka and 18 ka. During this time, steppe lions survived in western Asia, recolonizing Europe after the peak of the last glaciation. After these bottlenecks, *P. spelaea* apparently did reestablish a degree of genetic diversity millennia before it and other megafaunal species finally went extinct.

In an intriguing 2012 study, Marco Masseti and P. A. Mazza (both of the University of Florence) tackled another, related riddle. After steppe lions finally died out in Europe between 14,500 and 14,000 years ago they were replaced—after a second "lion gap" lasting about 6,000 years in eastern regions like Ukraine and Hungary—by the northward-dispersing savanna lion *Panthera leo*, at around 8 ka. By this time forests had replaced mammoth steppe in western Europe, and the ecological pressure of Late Paleolithic humans acted as another barrier. Reliably dated fossils from coastal sites in Italy and Spain, however, suggest that lions of some kind *did* exist in some peripheral Mediterranean areas during the so-called Lion Gap, but which species? One possibility, unsupported by any fossil evidence, is that *P. leo* could have entered eastern Europe before 8,000 years ago. The second is that late-surviving, relict populations of *P. spelaea* continued to exist in nonforested areas of eastern Europe until they finally became extinct during the early Holocene, preventing the dispersal of *P. leo* further west.

A fourth possible explanation for steppe lion extinction, as with the homotheres, is the specter of Late Paleolithic humans. Currently no evidence, apart from the possibly ceremonial practices mentioned in chapter 5, supports the idea that Aurignacians and other later cultures placed any significant hunting pressure on steppe lions. However, Clive Gamble (University of Southampton) and his associates in 2004 suggested that the warming climate and increasingly wooded ecosystems of Europe were conducive to human hunting and gathering. This may have led to an increase of human populations—and competition with pantherins—in regions like France and northern Europe beginning about 15.5 ka. Whether or not this was a factor, steppe lions were gone in Europe some 800 years later, and it's hard to avoid the conclusion that unsuccessful competition with human hunters as well as a decrease of suitable habitat and prey may have been decisive in their disappearance from the continent. The most recent reliable survival dates for *Panthera spelaea* in Europe are represented by a canine from Zigeunerfels, Sigmarigen, Germany, dated at 14,378 ka, and a skeleton from Le Closeau, northern France at 14,141 ka. The dating of steppe lions in eastern Beringia (Yukon, northwestern Canada) shows that they survived here about a thousand years longer, with the youngest date set at 13,290 ka. Beyond Beringia in southern Canada and the United States, the most recently surviving American lion *Panthera atrox* has a date of 12,877 ka for a specimen found at Consolidated Pit 48, Edmonton, Alberta. Modern humans (*Homo sapiens*) were probably not present in northern Asia before about 50 ka, but reliably dated human bone and organic-material artifacts show that they succeeded in dispersing first into western Siberia by 45 ka and then into Beringia by 40–30 ka. It isn't known if Beringia was continuously occupied or periodically abandoned by humans during the Wisconsin (Weichselian) glacial interval but pollen and faunal samples indicate that the south-central parts of the region during much of this time were ice-free refugia areas with a comparatively mild climate and abundant game. One human occupation site near the mouth of the Yana River dating to ~32 ka contained a wide spectrum of nonmarine animal bones, indicating that the inhabitants hunted interior game, and this and the types of domestic goods suggest that the Beringian occupations were sometimes extended—in which case, as in Europe and western Asia, there would have been conflict with steppe lions for open-country prey.

Farther to the south in midlatitude North America, populations of *Panthera atrox* may have been experiencing the same conditions: contraction of the Pleistocene prairies and along with it scarcity of their preferred prey, aggravated by competition from known pre-Clovis and other hunter-gatherer cultures. To some extent this fits with the "Pleistocene overkill" hypothesis, first proposed by Paul S. Martin (University of Arizona) in 1967. Martin believed, based on his extrapolations of increased human populations and rates of dispersal, that the postglacial extinctions of most of North America's megafauna—animals like horse, camel, mammoth, sloth, and other potential prey species—were caused by a lack of experience and time to adapt to the sudden appearance, relatively rapid radiation, and advanced hunting technology of immigrant Paleoindian cultures. These species were presumably unable to develop defensive strategies in coping with human predation, and their extinction in turn took with them dependent apex predators like American lion, sabertoothed cats, and short-faced bear (*Arctodus simus*). Some megafauna, such as mastodon (*Mammut americanum*), were already reduced to smaller, less contiguous populations by the climatic effects of the Holocene transition. Any additional ecological stress that increased mortality in such a slow-reproducing species could

FIGURE 6.8. Skull of *Panthera atrox* overgrown with grass.

cause local, regional, and finally continental extinction. As in the case of northern Asian ungulates there were certainly surviving prey species, such as modern bison (*Bison bison*) and elk (*Cervus canadensis*), that should have seemingly tided over *Panthera atrox* into the historic period, but, in the end, these were apparently not enough (figure 6.8).

CHAPTER 7
Man the Destroyer

Rome, 83 CE: On cue, the elevator box with its ingenious, groaning system of gears and pulleys slowly rises, lifting its frenzied burden, a Numidian leopard, to the sandy surface of the arena. Captured five weeks ago by locals who flushed it from a desert ravine, the leopard was driven into a pen, netted, and transported to ship, where it was thrown minimal amounts of meat to keep it hungry and prodded to keep it savage. Now, as it emerges from its underground dungeon into the blinding, early-morning glare on the Colosseum's floor, the cat's senses are assaulted by the terrible, swelling roar of awe and anticipation from a crowd of some 50,000 spectators. With perfumed, coiffeured, and hard-faced citizens looking on and grimy, spear-wielding *venatores* at the ready, the spectacle of slaughter is ready to begin (figure 7.1).

.   .   .   .   .

Twenty-three thousand years have passed since the time of the Aurignacians, and the great cats that once inspired awe and admiration have become impressive but disposable symbols of human power and mastery over nature. Beginning with the *venatio* (symbolic hunting or baiting) of wild Italian animals partly as a religious rite and partly to amuse crowds at the Circus Maximus as early as the third century BCE, the staging of wild-animal combats and deaths became ever grander, more expensive, and bloodier after the Roman Republic became the Empire. For earlier conquerors such as Sulla, Pompey, and Julius Caesar, *venationes* became an expected and traditional part of a *ludus*, or game, which they organized and paid for to celebrate a triumph, and for the later emperors, magistrates, and other officials, the costly annual event was a sure way of impressing the masses with their munificence and distracting them from civic problems. In addition to yearly games, special performances (*ludi extraordinarii*) might also be staged to accompany the dedication of a temple or other public building, and even a wealthy private citizen sometimes entertained the public with a scaled-down game

---

FIGURE 7.1. Entrance of Numidian leopard into the Colosseum, Rome.

FIGURE 7.2. Mosaic of Roman *venatore* in combat with leopard.
Source: Photo by A. Salvatore, courtesy of the Villa Borghese, Rome.

for a funeral or other personal occasion. If they had the influence, the citizens' agents and managers exerted pressure on provincials to organize and help pay in capturing animals for transport to Rome. Foreign animals, known widely as *Africanae bestiae* ("African beasts") regardless of their origin were probably rare in games (as opposed to local, European ones) before the beginning of the second century BCE. After this *bestiae dentate* ("beasts with [dangerous] teeth")—bears, big cats, and other large carnivorans—became increasingly common in the arena (figure 7.2). Followed later in the day by other events like gladiatorial combat, horse racing, athletic contests, and musical performances, *venationes* usually took place in the morning hours of the *ludi*. Obviously fierce and impressive species—whether herbivores like aurochs, bison, and rhinos or carnivorans like bears, wolves and big cats—made the most dramatic impression on the crowds, and these animals were forced to emerge into theatrically constructed shrubbery through side or underground entrances. As drums were beaten and trumpets were blown the confused and desperate animals were goaded into attacking spear-wielding *venatores* or javelin-wielding *iaculatores*, who dispatched them. Their deaths might be postponed long enough for a starved *besta dentatis* to serve as executioner for an army deserter, prisoner of war, or common criminal lashed to a post. Creatures commonly perceived by the Romans as natural rivals like boar and bear, hippos and crocodiles, and rhinos and bulls were paired up and, however unwilling, forced to fight each other to the death; even relatively gentle types like giraffes and elephants were sometimes destroyed for the sheer novelty of seeing how they might die by human hands.

For the big cats, it remained a curious paradox that while they were honored in some ancient civilizations as symbols of power and majesty, the animals were nevertheless slaughtered for royal sport and to symbolize an individual ruler's power over nature's most formidable creatures (plate 10).

After the conquest and destruction of Carthage in 146 BCE Rome annexed huge new areas that served as a source of leopards (*variae*, which included other spotted cats) and lions (*leones*), and the surrender of the eastern Seleucid Empire territories in 638 BCE made it possible to acquire tigers (*tigres*), which had formerly only appeared as rare and occasional gifts from that empire and the Indian kingdoms. The entire process of locating, catching, caring for, and transporting the big cats might take months when the source was as far away as Asia Minor (Syria, Iraq, etc.), but once an animal reached a Roman city, the final outcome was usually over in a few days. A typical show that took place in 169 BCE during the Roman Republic is described by the historian Livy, who wrote that at one *ludus* forty bears, some elephants, and sixty-three *Africanae bestiae*, in this case lions and leopards, were featured. To celebrate the election of his praetorship in 93 BCE the general Sulla arranged to have one hundred lions let loose in the arena, to probably attack one another before being killed by *iaculatores*. During the early years of the empire Augustus recorded that in a period of years encompassing twenty-six *venationes*, some 3,500 *Africanae bestiae* died, many of which were certainly big cats. Not to be outdone by his predecessor the Emperor Titus, for the dedication of the Colosseum in 80 CE, authorized one hundred days of celebration; on the opening day 5,000 animals are said to have been slaughtered, followed by 3,000 more over the next two days.

Long before they were sought for spectacles, however, big cats had been under human pressure since Neolithic times. Although lion and leopard populations in the lands surrounding the Mediterranean Sea, including North Africa and Asia Minor, were given some reprieve by the collapse of the Roman Empire in 476 CE, they continued to be indirectly threatened in other ways, beginning with habitat loss. Wood and charcoal were practically the only fuels used in the classical and early medieval worlds, and, over the millennia, incalculable amounts of them were used to provide heat, erect buildings, build ships, and make weapons. During Neolithic times the already fragile Mediterranean lands supported diverse ecosystems and fauna, but the widespread razing by the ancient civilizations of the cedar, oak, hawthorn, and walnut forests that had formerly ringed the

FIGURE 7.3. The destruction of the big cats' habitats in the Near and Middle East. As suggested by the now barren surroundings of this modern village in northeast Syria, ancient civilizations of the Near and Middle East relentlessly affected big cats directly by hunting and indirectly by eliminating their natural habitats. Once abundant in these regions and in western Asia, lions, leopards, and tigers could not survive when their ranges became increasingly fragmented. *Source:* Photo courtesy of UNESCO.

coastal areas altered rainfall patterns. Relentless plowing and replowing of land resulted in the windblown loss of soils, creating deserts where there had once been savannas, wetlands, forests, and open woods As a result of habitat loss the ranges of leopard and lion, which in early ancient history had formed a more or less continuous band from Turkey, Greece, and the Balkan countries into India, became seriously reduced and fragmented, and the remaining populations were faced with an ever smaller prey base as settlements expanded (figure 7.3).

Their environments having been denuded and rendered incapable of sustaining large ungulate populations, it was inevitable that pantherins would turn increasingly toward the easy prey of domestic cattle, sheep, and goats—setting them on a collision course with growing settlements. In the Mediterranean and adjoining regions, relentless human hunting—both for sport and in defense of domestic herds—for centuries had whittled away at leopards and lions. Whereas in some ancient Indian and northeast Asian traditional ethnic societies these big cats and their occasional predations were tolerated as an integral part of the order of the natural and human cosmos, the pastoral cultures of the Middle and Near East viewed pantherins and other carnivorans as dark entities to be feared and eliminated. Following the collapse of the classical world, Middle Eastern peoples experienced widespread migration, disease, and war, as well as changing political, religious, and social orders. New identities were forming and, for poor, agricultural classes, any reverence or appreciation for nature was generally left behind in the hardscrabble struggle for survival from year to year. The loss of an ox, cow, or donkey could mean severe hardship or even economic catastrophe for a family. The wilderness that lay beyond the fields was to be mistrusted, if not feared outright, and all carnivorans were now enemies. By 100 BCE Asiatic lions were mostly gone from Mesopotamia, northern Greece, and Macedonia but survived in Palestine until about the time of the Crusades, 1095–1291 CE; small pockets hung on in Syria, Iraq, and Iran until well into the twentieth century. In spite of their gradual decline and extinction in these areas lions, leopards, and cheetahs were still common in savanna and open woodlands in sub-Saharan Africa as late as the end of the nineteenth century. In western Asia smaller numbers of the same three cat species shared the region, although not the same habitats, with Caspian tigers. Unlike those of Europe and the Near and Middle East, India, and China the natural ecosystems in most of the rest of western Asia were still largely intact and unspoiled before historic times because of smaller population growth, high human mortality, and less intensive land use.

The comparative refuge these regions provided for big cats and other wildlife abruptly changed with the advent of Western colonialism and modern guns. As in previous centuries hunting was, as it had always been, a common recreation for historic ruling classes such as the Assyrians, the Moghuls, and much later royals who thought nothing of killing huge numbers of animals as status symbols during a single hunt (figure 7.4). For occupying Western colonialists, the faunal richness of Africa and Asia was considered to be an inexhaustible source of game for recreational hunting, and the terrible efficiency of European rifles made it possible to kill more animals than ever before. This was the mark of a hunter's status, and no kill rated more highly than a big cat. In colonial India "the bag" was a major preoccupation not only for maharajahs, British military brass, and civil authorities but also for many expatriate civil servants during the long hill-station summers. Hundreds of tigers, leopards, and lions were shot every year in India and Africa, with no consideration as to age or gender (figure 7.5). The largest and most impressive were removed from the gene pool to end up as trophies on walls. As early as the 1870s, however, in parts of colonial India some hunters began to notice a decline in game, with J. Baldwin remarking in 1877, "In former years tigers were doubtless a scourge, now they are becoming rare even in the wooded parts of the country. . . . Where once a dozen could be shot by a party . . . two or three will now only be bagged" (quoted by Schaller in *The Deer and the Tiger: a Study of Wildlife in India*).

FIGURE 7.4. Nineteenth-century British hunting party with bag of tiger corpses. *Source:* Photographer unknown. Courtesy of Panthera.

FIGURE 7.5. Regimental photo of young British officers and their trophies. *Source:* Photographer unknown. Courtesy of Panthera.

Probably because of India's smaller geographic size and larger rural human populations, the decline in big cats and wildlife in general was at first more apparent there than in Africa, and acts by the British authorities, although only nominal, were passed to protect animals from excessive shooting in 1887, 1912, and 1935. Later, wildlife reserves including the Banjar Valley Reserve, Corbett National Park, and Kaziranga Sanctuary were established but, part because of governmental disinterest in controlling hunting, pantherins still declined. In spite of this, India retained much of its wildlife until 1947 when independence brought with it a wholesale slaughter, in part a reaction to colonial repression. In response to food shortages, guns were issued to farmers, who shot huge numbers of wild ungulates like blackbuck (*Antilope cervicapra*) as crop pests and domestic-animal competitors, bringing a formerly common species to the point of extermination. Indiscriminate night hunting from vehicles became common. At the same time, habitat destruction intensified with population expansion, domestic animals' overgrazing, exhaustive plowing, and forest cutting for cultivation and railroad ties, rendering once virgin grasslands, forest, and jungle into arid tracts of thorn woodland and desert. As a result the Indian subcontinent has lost about 75 percent of its native forests, much of it tiger and leopard habitat (figures 7.6, 7.7). By the 1950s the national and state governments, aided by private and provincial conservation groups, slowly began to address the issue, establishing a number of small but significant reserves that, although still beset by poaching, wood gathering, and other illegal activities, are still the nuclei for protecting tigers, leopards, lions, and other remnants of India's once rich natural heritage. Whether they come from India or not, tigers have for millennia had the unhappy distinction of having body parts that are falsely considered antidotes to erectile dysfunction and other human failings. Any part of a tiger is avidly sought after, and extensive black markets still operate to supply these (see chapter 8).

Partly because of its vast extent and partly because of a more enlightened colonial administration, the largest and most diverse savanna and open woodlands of Africa fortunately received reasonably effective if uneven protection beginning in the late nineteenth century. Africa's first national park was established in 1925 when Albert I of Belgium designated an area of the Virunga Mountains as the Albert National Park (since renamed Virunga National Park) in what is now Democratic Republic of Congo. In 1926, the government of South Africa designated Kruger National Park as that nation's first, although it was an expansion of the earlier Sabie Game Reserve established in 1898 by President Paul Kruger to forestall the excessive hunting of lions. The Serengeti National Park in Tanzania followed in 1951. By the mid-twentieth century there was a major cultural shift; localities that had long been popular hunting destinations now became tourist attractions for photographic safaris. By the time the sub-Saharan nations

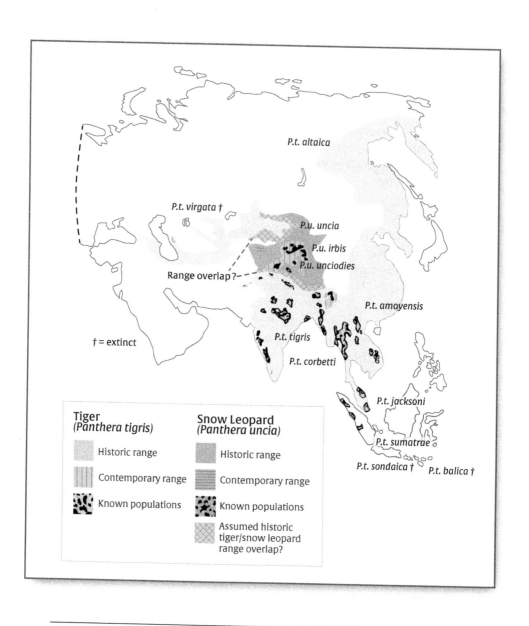

FIGURE 7.6. Map of Asia, showing historic and contemporary tiger/snow leopard range.

Source: Modified from Wheeler and Jefferson (2009).

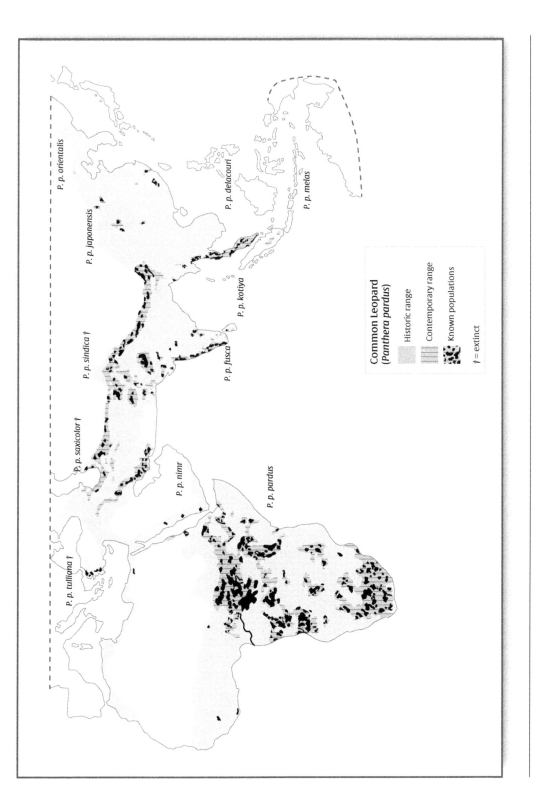

FIGURE 7-7. Map of Africa/Eurasia, showing historic and contemporary common leopard range. *Source:* Modified from Panthera (2009).

gained independence in the late 1950s and 1960s, the value of the game parks to national economies was well recognized, and their importance was later ingrained into the awareness of Western cultures through popular books and films. Today, habitat loss, the invasion of wild areas for domestic animal grazing, large-scale killing for "bush meat," and foreign cultural demand for rhino horn and elephant ivory are the most immediate threats to African wildlife, followed by pressure from agriculturalists to have access to natural land and by militarists who profit from large-scale, organized poaching. While they are in the game parks, lions and leopards are reasonably secure but are subject to shooting and poisoning when they venture outside the borders, which has catastrophically decreased their range (figure 7.8). All of the African cat species are classed as vulnerable by the International Union for the Conservation of Nature (IUCN), not only from outright killing and range shrinkage but also because, as with other surviving pantherins, habitat fragmentation results in a population's genetic isolation. Small and isolated populations result in a decrease in variability from inbreeding, and the species becomes less resilient to disease and less able to adapt to change.

Although less well publicized than their counterparts in Africa and India, the ecological richness and wildlife diversity of Central Asia are today under an equally serious threat. As late as the 1920s and 1930s, regions like Tibet and Mongolia supported huge herds of chiru or Tibetan antelope, gazelle, kiang or wild ass, yak, and Bactrian camel on the open plains while white-lipped deer, argali, and bharal flourished in the hills and high precipices. Some of these were staple prey for common and snow leopards. Although for decades Central Asia was less accessible than other parts of the world because of the turmoil of war and changing politics, it was soon discovered to be a hunter's paradise, and the losses began. Long after the lions and tigers had been killed off, sports and subsistence hunting of ungulates in a largely poor, pastoral culture began to take its toll; this was compounded by the uncontrolled slaughter of Tibetan antelope or chiru for their fine, fleecy wool, known as *shatoosh*, reaching its height in the 1970s and 1980s. Like the killing of native snow and common leopards, whose spotted coats fed a clothing fashion trend, chiru killing still continues, and the wool is now turned into scarves in Nepal as well as Kashmir. Chiru and other native ungulates still continue to suffer from sharing their fragile grazing areas with hordes of sheep, goats, and domestic yak. Snow and common leopards, whose large ranges are in inverse proportion to their prey availability, may turn to taking livestock in order to survive, and, as in Africa and India, they are shot or poisoned by herders and farmers who are justifiably embittered by losing their animals to a predator. Snow leopards, whose pelts continue to be sold illicitly by locals, are shot and poisoned as livestock predators and have, like other Asian big cats, suffered a reduction in range and numbers.

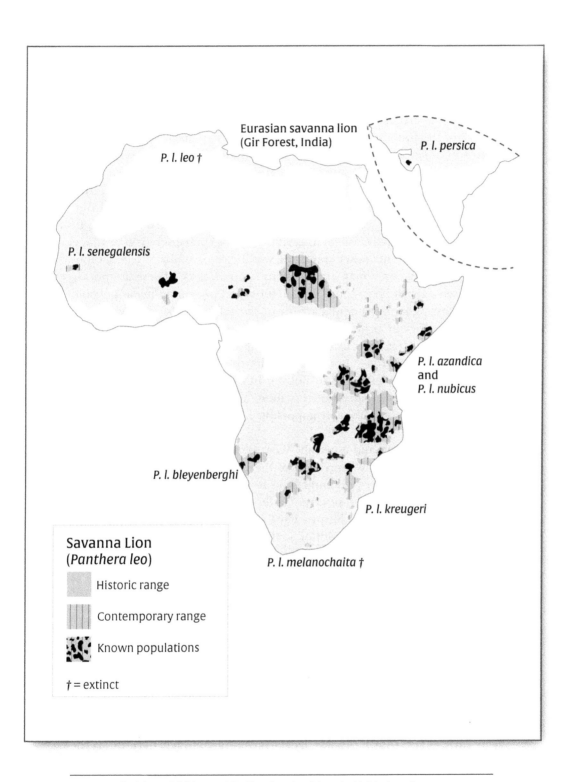

FIGURE 7.8. Map of Africa and India, showing historic (Africa) and contemporary savanna lion range (for complete historic savanna lion range, see fig. 6.6). *Source:* Modified from Panthera (2014); Gujarat National Parks (2017).

The gallery-forested riverbanks and tropical rainforests of South America, the stronghold of jaguars and other diverse non-pantherin cats, were in the spotlight during the 1930s as a source of spotted fur for trimming purses, stoles, and coats. The focus continued in the 1950s as a source of smaller species for the exotic pet trade. Countless cats were killed for this in spite of the fact that, in contrast to pelts of wolverine, bear, fox, and other arctoids, the fur of most felines is a less effective insulator and soon "slips," or falls out, with wear. However, the main threat today to jaguars and their other South American relatives is habitat loss, in this case from the widespread destruction of rainforest for commercial timbering and to create vast cattle ranches to feed the Western addiction to beef. This is especially critical in Brazil, where 48 percent of jaguar range is located. As in some other countries, wildlife reserves that are geographically impressive on a map are in reality "paper parks." These are poorly protected against illegal land exploitation and poaching, made possible in part through governmental apathy or outright payoffs. The hunting of jaguars is officially prohibited in nations such as like Argentina, Belize, Colombia, Nicaragua, and the United States. Others—such as Mexico, Costa Rica, and Peru—allow killing under the guise of predator control. Trophy hunting is still legal in Bolivia. In Ecuador and Guyana, jaguars have no protection and, as of this time, an accurate assessment of the number of animals is probably impossible because of their continent-wide range (figure 7.9).

All African lions, and especially West African populations, are listed as vulnerable by the IUCN, and most of the threat is caused by agricultural encroachment and the poisoning and shooting of lions by pastoralists outside of national parks and reserves. Sadly, some lion subspecies, like other pantherins, now belong to history. The Cape lion, *Panthera leo melanochaita*, whose range occupied the velds of South Africa, is totally extinct as a result of uncontrolled hunting. Like Barbary lions this was a very large subspecies (males approached 270 kg [600 lb.]), distinguished by a particularly long, luxuriant black mane, fringes on the flanks and belly, and a tawny facial fringe. Christiansen (2007) and other zoologist-paleontologists have recently described the skull as being longer and narrower, with a deeper muzzle and shorter postorbital region, than in other African lions. The rapidity with which Cape lions became extinct after European contact in the early nineteenth century suggests that hunting by Dutch and British settlers and sportsmen was the cause rather than habitat loss. The last known wild individual was shot in 1858.

The Barbary or Atlas lion (*Panthera leo leo*), probably the largest of the African lion subspecies was, by classical times, already hard hit by the demand for it as a spectacle in the arenas. Its distinctive appearance made these lions a desirable target for capture by Roman agents, who sometimes

FIGURE 7.9. Map of North America and South America, showing historic and contemporary jaguar range.
Source: Modified from Panthera (2009).

FIGURE 7.10. Barbary lion in captivity.
Although extinct in the wild, if a large and diverse enough gene pool exists in
zoo-bred captives the subspecies could possibly either be back-bred or genetically
reengineered. *Source:* Photographer unknown. Courtesy of the Royal Zoological Society.

termed them *archileontes*, or "ruling lions." In the wild they were characterized not only by exceptional size (for males, 2.3–2.8 m [93–110 in.] in
length and 270–300 kg [600–660 lb.] in weight) but also by long, dark, and
luxuriant manes, which often as in Cape lions extended posteriorly on a
distinctive fold of skin along the sides of the belly (figure 7.10). Barbary
lions formerly inhabited the North African regions that included the Atlas
Mountains but, under human pressure and like the surviving Asian lions,
this subspecies had extended its range from plains habitats to wooded hills,
where it preyed on Barbary red deer (*Cervus elaphus barbarus*), Cuvier's
gazelle (*Gazella cuvieri*), and Eurasian wild pig, among others. Despite
modern guns and habitat loss, Barbary lions continued to persist in small
numbers until the middle of the twentieth century. The last known wild
individual was shot in 1942 in Morocco, but small numbers may have
existed there and in Algeria until the 1960s. In spite of their extinction in
the wild, isolated zoo-bred lineages that mostly or entirely have Barbary
lion genomes still survive, and the WildLink International program seeks
to raise, habituate, and release back-bred individuals into their native
Atlas Mountains.

There are five other generally extant, recognized subspecies of
African lions: the Congo lion (***Panthera leo azandica***), northeastern Democratic Republic of Congo; the Angolan lion (***P. l. bleyenberghi***), southern

Democratic Republic of Congo and neighboring parts of Zambia and Angola; the Transvaal lion (**P. l. kruegeri**), Kalahari region to eastern South Africa; the Nubian lion (**P. l. nubicus**), east and northeast Africa; and the Senegalese lion (**P. l. senegalensis**), West Africa east to the Central African Republic. Although still surviving, all have collectively gone through major population declines of 30–50 percent since the early 1970s. Much of this was caused by habitat loss from the same forces of agricultural encroachment. All, especially the West African populations, are now listed as "vulnerable" by the IUCN.

Asiatic, Indian, or Persian lions (***Panthera leo persica***), once widespread over vast areas of eastern Europe and western Asia, have suffered the loss of almost all of their range and now survive only in the thorn-woodland areas of the Gir National Park and Wildlife Sanctuary (Sasan Gir Conservation Area) in Gujarat, India (figure 7.11). By the mid-nineteenth century lions had mostly disappeared in the subcontinent, with one shot in the state of Madhya Pradesh as late as 1851. Male Asiatic lions typically weigh 160–190 kg (350–420 lb.) and have a distinctive grayish-buff color in their

FIGURE 7.11. Asiatic lion.
The once vast geographic range of the Asiatic lion (*Panthera leo persica*) has been reduced to a small forested area on the Gujarat Peninsula of northwestern India. Males, pictured here, typically have much less extensive manes than those of African lions and may have a distinctively gray overall body coloration. *Source:* Photo by Ravindra Singh. Courtesy of Bhadra Wildlife Conservation Society, India.

coats, a larger tail tuft than African lions, and extremely sparse manes in males, allowing the ears to show. Both sexes have a distinctive longitudinal fold along the belly that resembles that of Cape and Barbary lions, and their skull differs from that of African lions in that the infraorbital foramen of the skull is divided by a septum and the auditory bullae are less inflated. Smaller pride sizes and less sociability compared to African lions are adaptations to the smaller sizes of prey now available to them, and their thorn-forest habitat is now all that is left following centuries of human encroachment.

Of the nine generally recognized subspecies of tigers that survived into historic times, three are now extinct because of hunting and habitat loss. The first to be lost was the Balinese tiger (***Panthera tigris balica***), native to the island of Bali in Indonesia. This was the smallest tiger, with males usually reaching lengths of 1.6 m (63 in.) and weights of 90–100 kg (200–220 lb.), and its size was a response to the limited prey resources of its small island, which included Javan banteng (*Bos javanicus javanicus*) and common (Sumatran) muntjac (*Muntiacus montanus*). Like other subtropical and tropical tigers it had short, dense fur and a bright and reddish upper-body color but had a purer white undercoat than the other subspecies (figure 7.12a). Around 1912, Balinese tigers were still considered common throughout the island, but this rapidly changed in the interval between the two world wars, when uncontrolled hunting by the Dutch colonials and by armed ethnic locals caused the tigers' numbers to plummet. By the middle of the 1930s a survey of the Dutch Indies by the recently formed International Wildlife Protection Commission reported, " A few yet live in West Bali but they are having a hard time because they are much sought after by[colonial] hunters from Java, so they will certainly disappear within a few years. The species [*sic*] also exists in northwest and southwest Bali" (Day 1981). This was a perceptive statement, and the last wild Balinese tiger, a female, was shot at Sumbar Kima, west Bali in 1937. No known genomes are preserved in any zoo-bred populations.

This extinction was followed sometime during the late 1970s by that of the Javan tiger (***P. t. sondaica***), a similarly geographically restricted subspecies with a coat pattern much like that of the Balinese but that larger on average, with weights of 100–141 kg (220–311 lb.) in males (figure 7.12b). As with the Balinese tiger, poorly restricted hunting determined its fate, and, on the larger but heavily populated island, the loss of forest to timbering and plantations contributed. Although "paper park" sanctuaries in some areas were in existence by this time, the last confirmed sighting for a Javan was at the region of Mt. Betiri in 1979, and a later effort in 1990 to find any traces of the tiger was a total failure.

The Caspian, also known as the Hyrcanian or Turan tiger (***P. t. virgata***) apparently lasted until the final decades of the twentieth century. Caspian tigers, whose **genome** was very similar to the surviving Amur (Siberian)

(a)

(b)

FIGURE 7.12. Extinct Balinese and Javan tigers.
(a) The skin of a Balinese tiger *Panthera tigris balica*. Like other tropical tiger races, this small subspecies had a bright, coppery dorsal coat and a comparatively pure white undercoat. Found only on the island of Bali, it had the smallest range of all tiger subspecies and was therefore the most vulnerable to extinction. (b) One of the last Javan tigers (*Panthera tigris sondaica*), shot during the 1930s. *Source*: Photographer unknown. Courtesy of Panthera.

FIGURE 7.13. Caspian tiger in captivity.
This photo (of unknown date) of a male zoo specimen displays the typically long belly fur and cartoonlike side whiskers that made these cats so distinctive. *Source*: Photographer unknown. Courtesy of the Royal Zoological Society, London.

tiger (*P. t. altaica*) of Manchuria and eastern Russia, were large at 3 m (10 ft.) in length and 170–240 kg (370–530 lb.) for males, intermediate in size between the Amur and Indian subspecies. Their appearance was highly distinctive, with long, drooping facial side ruffs and long belly fur as well as pale coats (figure 7.13). They inhabited the open woodlands, marshes, and more heavily vegetated river corridors east, west (to eastern Anatolia, Turkey), and south of the Caspian and Black Seas and across the Pontic-Caspian steppe into Central Asia (where they were probably sympatric with western Asian lions). They were also occasionally sighted in riparian locations of the "cold desert" of Takla-Makan in Xinjiang Province, China. In spite of such a widespread range and adaptability to diverse habitats, Caspian tigers began to decline in the late nineteenth century, traceable to the Russian colonization of Turkestan at that time. As happened with the declines of other big cats, the importation of modern, fast-loading rifles made it easy to shoot not only tigers but also one of their staple prey species, the Eurasian wild pig. These underwent a rapid decline not only from overhunting, beginning in the mid-nineteenth century through the 1930s, but also from natural disasters and epidemics of swine fever and foot-and-mouth disease. The regular Russian army was also used to eliminate predators from around woodlands, farms, and other settlements, and, until World War I, as many as one hundred tigers per year were shot in the forests around the Amu-Darya and Piandj Rivers, with high bounties paid per skin. A second major factor was habitat loss. Although extensive, the Caspian tiger's riparian and reed-bed habitats, where the largest populations were concentrated, were often patchy and separated from one other, and this worked against their survival when marshes, prized for their rich silt, were drained to grow cotton and other crops. Intensifying through each successive decade of the twentieth century, the destruction of tiger habitat, in combination with predator control and sport hunting, resulted in the cats' gradual decline over their formerly large range. Areas like the reed beds of the Ili River in Kazakstan, the Sumbar Valley of Turkistan's Kopet-Dag mountain range, and the river drainage into the Aral Sea were among their final strongholds. The last confirmed sightings and kills were scattered across the 1960s, the mid-1980s, and even possibly as late as the early 1990s, after which they ceased.

Until recently, many taxonomists had traditionally recognized six extant subspecies of tigers based on skeletal morphology, body size, coat character, and geographic distribution (see chapter 4). These were the Amur (or Siberian) tiger (*Panthera tigris altaica*), the Indian or Bengal tiger (*P. t. tigris*), the South China tiger (*P. t. amoyensis*), the Malayan tiger (*P. t. jacksoni*), the Indo-Chinese tiger (*P. t. corbetti*), and the Sumatran tiger (*P. t. sumatrae*). However, in a 2015 study based on genetics Andreas Wilting (Leibnitz Institute for Zoo and Wildlife Research) and his associates demonstrated that extant and recently extinct tigers fall into two main evolutionary groups, the continental lineage and the Sunda lineage. They accordingly proposed the recognition of only two subspecies: *Panthera tigris tigris* (Bengal, Malayan, Indochinese, South Chinese, Amur, and Caspian populations) and *P. t. sondaica* (Javan, Balinese, and Sumatran populations). This has great value in the allocation of resources of future captive-breeding and rewilding programs. Although adapted to habitats that now range from dense Siberian taiga forest to open temperate woodlands and tropical mangrove swamps, these big cats have lost 93 percent of their former range. The majority of wild tigers now live in India in small, fragmented national parks and reserves. The entire global wild population is estimated by the IUCN at between 3,062 and 3,948 individuals, and the species as a whole is considered endangered.

Leopards, in spite of their still enormous overall range, opportunistic hunting behavior, extreme adaptability to different habitats, and lower profiles, have fared only little better than their bigger relatives. Two subspecies, the Anatolian leopard (*Panthera pardus tulliana*), once native to western Turkey, and the Balochistan leopard (*P. p. sindica*), which lived in southern Iran, southern Afghanistan, and southwestern Pakistan, became extinct around the 1970s. Others are now gone from such areas as eastern coastal China, Korea, and parts of North Africa and the Near East, where they were once common. Of the nine surviving subspecies recognized by the IUCN, all have suffered extensive habitat loss and fragmented populations, with the Amur, Far Eastern, or Siberian leopard (*P. p. orientalis*) and Javan leopard (*P. p. melas*) considered to be the most critically endangered cats in the world. The other surviving subspecies include the African leopard (*P. p. pardus*), the type subspecies from Africa; the Arabian leopard (*P. p. nimr*); the Persian leopard (*P. p. saxicolor*), from Central Asia; the Sri Lankan leopard (*P. p. kotiya*); the Indian leopard (*P. p. fusca*); the Indo-Chinese leopard (*P. p. delacouri*), from southeast Asia into southern China; and the north Chinese leopard (*P. p. japonensis*).

Of the eight subspecies of jaguar recognized by Kevin Seymour (Royal Ontario Museum) in his 1989 summation of the jaguar for the *Mammalian Species* series of the American Society of Mammalogists, only one, the Arizona jaguar (**Panthera onca arizonensis**) is no longer extant. The last documented killings and sightings of Arizona jaguars, distinguished by their size and flatter, more depressed nasal bones, were in 1904 and 1905. Recent confirmed sightings in Arizona of a big male jaguar known as "El Jefe" and of others by wildlife biologists within the last few years are more likely to be dispersing representatives of another subspecies in the Southwest, the western Mexican jaguar (**P. o. hernandesii**). This and another subspecies, the southeastern Mexican jaguar (**P. o. veraecrucis**) are also threatened, hovering near extinction since the beginning of the twentieth century. The other named extant subspecies are *P. o. onca*, *P. o. centralis*, *P. o. goldmani*, *P. o. peraguensis*, and *P. o. peruviano*. The IUCN doesn't recognize any formal subspecies, as these don't appear to be supported by genetic or morphological evidence. It does, however, recognize four incompletely isolated physiogeographic groups—Mexico plus Guatamala; southern Central America; northern South America; and southern South America—but no formal taxonomic names are assigned to these populations. Like their lion counterparts in the European Paleolithic, jaguars have slid from being revered as gods by the Aztecs and Mayas to the status of vermin, subject to shooting by ranchers and farmers. Beginning in the 1960s jaguar populations began an intense decline from alleged predator control, the skin trade, and steadily fragmenting habitats, especially with the razing of tropical rain forest for cattle ranches and plantations. Jaguars now occupy approximately less than half of their historic range.

Panthera uncia, the snow leopard or ounce, derived some natural protection until the mid-twentieth century by its remoteness from most human contact. To date less than 2 percent of its known range has been mapped. Because of their particularly harsh, unforgiving habitats, snow leopards have probably never been as common as the other great cats and have suffered serious losses over the last few decades from hunting for their pelts. Although naturalists in the 1960s considered the possibility that more than one subspecies of these cats existed across their historic range of China, Central Asia, Mongolia, and India, a recent DNA fecal analysis by Jan Janecka (Duquesne University), Rodney Jackson (Snow Leopard Conservancy), and their associates confirms that there are *three* distinct races, geographically separated into northern, central, and western regions. These are **P. u. uncia**, inhabiting Central Asia, including the Alay, Karakoram, Pamir, Tian Shan, and trans-Himalayan mountain ranges; **P. u. unciodies**, present in the Tibetan Plateau and core Himalayan mountain ranges; and **P. u. irbis**, found in the southern Gobi Desert and Altai Mountains of Mongolia. The genetic studies that established the three subspecies also indicate that the cats seem to have undergone a genetic bottleneck about 8,000 years ago during a time known as the Holocene Climatic Optimum, when the earth's temperatures temporarily rose. This likely reduced the distribution of snow leopard populations, reflecting the species' vulnerability to higher regional temperatures. Until recent decades no field studies had been undertaken of this cat and, as a result, estimates of the world's snow leopard population in the wild vary considerably, from 4,500 and 10,000. The number is thought to have fallen by 20 percent within the last sixteen years. As global warming makes the formerly colder higher altitudes more attractive to human settlement than during historic times, encroaching pastoralism in these now available areas makes conflicts with herders ever more likely, with poisoning and shooting, as in the past, being common responses to livestock losses.

The human destruction of animals once revered for their power and beauty has been immense.

CHAPTER 8
The Steeds of Durga

Sikhote-Alin Nature Reserve, Primorsky Krai, Russia, 2021: As they awkwardly clamber across a swift, frigid stream, a team of Russian and American wildlife biologists is electrified by a sudden series of staccato pips. The tracking device, emitting slow beeps for a day and a half, has once more come to life, indicating that a young female Amur tiger, "Durga," is again active. Earlier in the year Durga was captured in a baited box trap at the edge of a clearing, where the 250-pound cat was anesthetized with Ketaset injected into her haunch. As she lapsed into grogginess, Valium was administered to reduce anxiety and muscle tension as her collar, fitted with a transmitting radio, was attached. The variable frequencies, one for each individual tiger in the study area, are programmed to reveal changing behavior patterns: slow and steady when the animal rests, quicker and louder when there is body movement, and going from loud-to-soft according to the terrain over which the collared tiger travels. Now recaptured after a few months, Durga gives the researchers vital information about her movement, her activity patterns, the size of her home area, and her preferred habitats. This will help to form a database of information from which the size of a new peripheral area for the small but expanding tiger population can be assessed. Now almost fourteen months old, Durga has left her mother and made contact with a huge male, "Ivan," who has regularly left imprints in the snow and sprayed pungent urine on the moss-covered bark of the forest's oak and linden trees. His sonorous moans and persistence in following her ensured that they would come together again, and it's likely that the silence of their collars was merely a respite in the tigers' courtship. If their mating is successful, it adds to the hope that Amur tigers will persist in the future (figure 8.1).

. . . . .

Information about big cats in the wild, although now considerably helped by radio telemetry and other technologies, is usually hard won and is as often attended by failures and heartbreak as by success. Field observations of the behaviors of a tiger, snow leopard, or other pantherin are the raw material from which meaningful and realistic assessments of the cats' needs are derived, but these must then be translated into actions on a human level. This means rigorously enforced management to give the animals the space they need to hunt and reproduce, to protect their lives and habitats from destruction through commercial greed, and to prevent misguided and needless killing from the humans who share their world. Cultural and political realities in the countries where big cat habitats are located must be addressed and, ultimately, they determine if a species will truly survive.

The earliest efforts to protect big cats were spurred by the realization that uncontrolled shooting would soon lead to their extinction in the wild. These were sometimes initiated by responsible professional hunters such as James ("Jim") Corbett and Kenneth Anderson in colonial India who, in spite of hunting "man-eating" tigers and leopards for a living, respected the animals; they advocated for laws to establish reserves and sanctuaries

FIGURE 8.1. Future research team radio-collaring an Amur tiger.

for their protection and to halt excessive hunting as well as the devastation that occurred from the progressive elimination of their habitat. Pressure to create these was added by conservation societies. Although well-intentioned, the perimeters of the reserves were often mapped out on a cursory and intuitive basis and lacked any actual field studies to determine the behavior, ranges, and other requirements of the big cats and their prey species. This is exemplified by what later became the Serengeti National Park, Tanzania, in which the true extent of the migration routes of some 1.5 million western white-bearded wildebeest (*Connochaetes taurinus mearnsi*) and 250,000 Burchell's zebra (*Equus quagga burchelli*) were originally underestimated when the park was established in 1951. The boundaries were later extended by the addition of nearby natural reserve areas, but this error pointed to the need for long-term scientific studies, rather than sporadic observations, to serve as a basis for a greater understanding of life cycles, behaviors, and habitat needs that are essential in maintaining the health of both predator and prey species.

In the Serengeti the long-term management and protection of the iconic game herds demanded an assessment of the effects on these of the lions and other important predators, prompting the beginning of groundbreaking studies such as Schaller's 1972 work, *The Serengeti Lion: A Study of Predator-Prey Relations* (figure 8.2). In the late 1960s Schaller had tackled

FIGURE 8.2. George Schaller observing African lions in Serengeti National Park, Tanzania c. 1968.
Source: Photo by Kay Schaller.

the complex ecology of tigers and ungulates in Kanha Park, Madhya Pradesh State, India—a limited but relatively unspoiled area from which the animals' ecological needs could be determined—which resulted in *The Deer and the Tiger: A Study of Wildlife in India* (1967). This led to a better understanding of the requirements of the feline predators and their prey, a solid database from which the optimum extent of reserves could be best determined, and guidelines as to which wildlife-management practices in these and surrounding areas would ensure the health of the animal populations. In making studies like these, Schaller took into account not only the natural dynamics of the animals but also their relationship to the human communities living in and around the park, as did Alan Rabinowitz (Wildlife Conservation Society), a naturalist who studied jaguars in Belize's Cockscomb Basin in the early 1980s (figure 8.3) and, later, wild cats in Thailand's Huai Kha Khaeng Valley.

FIGURE 8.3. Alan Rabinowitz studying jaguar in the Cockscomb Basin, Belize, early 1970s.
Source: Photo by Steve Winter, *National Geographic.* Courtesy of Panthera.

The most immediate danger to pantherin (and small cat) survival from local humans is poaching. In the past, hunting of wild "bush meat" was always a way of obtaining needed animal protein for both agricultural and pastoral societies, but now sustenance hunting has given way to greed and disregard for a traditional relationship with nature. The advent of modern guns and government-funded roads, projects that lead to more contact with urban consumer values, and the growing demand for exotic animal products now make killing native animals an attractive economic incentive for rural people who live close to wildlife. In spite of longstanding international guidelines, laws, and sanctions such as the United Nations Convention on International Trade in Endangered Species (CITES) and the rules of the Association of Southeast Asian Nations (ASEAN), poaching is still the most common and immediate threat to wildlife. For big cats this mainly takes the form of killing for the body-parts trade. The market for pantherin parts has traditionally been in China and Southeast Asian countries like Thailand, Burma, and Laos, where animal organs have for centuries been credited with imagined medicinal benefits. Tiger bone, powdered as pills or soaked in rice wine, was until recently considered by some to cure rheumatism and arthritis, and tiger meat supposedly gave strength (figure 8.4). The Chatuchak weekend market in Bangkok is a sprawling, open-air "shop of horrors" where tourists can purchase clouded leopard, snow leopard, and jungle cat amulets and pelts. Wealthy men,

FIGURE 8.4. Tiger and other pantherin body parts for sale.
Source: Photo by Li Quan.

often business people from China, can attempt to restore their flagging virility by quaffing US$25 whiskey shots flavored with a tiger's heart or spend US$300 for a takeaway carton of hot tiger-penis soup. From hubs like the Suvarnabhumi International Airport in Bangkok, dealers smuggle all kinds of exotic animals, both dead as potential souvenirs and barely alive for the pet trade, via intermediaries in transit spots like Cambodia and Morocco. These are hidden in legitimate products and clandestinely sent to buyers in Saudi Arabia, Japan, Europe, and the United States. The animal trade is supplied either by the trafficker's agents, who pay local rural inhabitants to poach, or by hired outside gangs. As the money offered is often more than an average rural person can hope to make over a long period of time, there is always an incentive to keep poaching. In addition to the souvenir trade there is also a hugely profitable market for décor and foods from prestigious animals—exotic, desirable, and therefore expensive. This is the kind of business epitomized by ventures like the Kings Romans Group, a Hong Kong–based company that signed a lease with the Laotian government for a twelve-square-mile commercial plot across the Mekong River from Thailand. A schizophrenic blend of hotel, casino, shopping center, cock- and bullfighting ring, and ill-kept zoo, it's run by and caters mainly to rich Chinese. The "zoo" is suspected as a source of exotic meats for the restaurant's "special jungle menu," which offers plates of sautéed tiger for US$45. Nearby, a number of jewelry and pharmacy shops sell high-priced tiger teeth and claws as well as rhino-horn carvings and elephant skin and ivory. Kings Romans is but one of several operations profiting from endangered animal parts, openly sold in the so-called Golden Economic Zone in the Bokeo Province of Laos.

In response to market demand and declining supply of wild tigers, it was perhaps inevitable that entrepreneurs during the last few decades realized a profitable but hideous alternative: tiger farms (figure 8.5, plate 11a, b). In 2016 the "Tiger Temple," advertised as a combined religious shrine and petting zoo run by local Buddhist monks and intended for education about and appreciation of these cats, was raided by Thai officials. In addition to more than one hundred living adult tigers they found the bodies of many cubs, some frozen and others preserved in alcohol, waiting to be processed for souvenirs. One monk, arrested while attempting to escape, was transporting tiger pelts, teeth, and over a thousand amulets featuring tiger skin. Tiger farms have mushroomed as small- to industrial-scale operations in Thailand, China, Laos, and Cambodia, and although sometimes masquerading as zoos, sanctuaries, or even "conservation breeding centers," they are in reality simply factories for the breeding and slaughter of tigers for their body parts. Cubs, like adults, are a tourist draw, but at an early age these are quickly taken away from their mothers, forcing the females back into heat so they may breed again as quickly as possible. Just like cattle, adult tigers are killed and butchered, and their meat, bones, and

(a)

(b)

FIGURE 8.5. Tiger farm in China.
As of this writing, the Chinese government's State Forestry Division not only
sanctions but also invests in and promotes the private operation of tiger breeding
facilities, where the animals are kept in overcrowded and miserable conditions
(a) until they are slaughtered for their body parts (b).

other body parts are processed for sale in high-end food and luxury-item markets. The less valuable left-over parts are used for crafting tourist items.

Some have claimed that such slaughterhouses take the pressure off wild tigers, but in reality they only *increase* the demand for wild tiger parts by making these, by virtue of their comparative rarity, appear superior and therefore more prestigious and valuable than those of captive-bred animals. And, as with investors in gold or rare art, there are those who currently and secretly invest in the tiger-parts trade, among them the State Forestry Division (SFD) of the People's Republic of China. As incredible as it sounds in light of China's official 1993 law banning tiger imports, both poached and farmed, from other countries, the SFD not only sanctions but actively promotes Chinese domestic tiger farming, as well as wineries on tiger farms.

This was observed and documented by author and wildlife activist Judith Mills in her 2015 book, *Blood of the Tiger*. The contradiction comes from a Chinese legal loophole: the 1993 ban (which is now superseded, as we discuss later), while specifically criminalizing the importation of wild-cat body parts, actually allowed the harvesting of tiger and other endangered-species products derived from animals *domestically farmed in China*; it was crafted to bypass and supersede Chinese wildlife-protection laws that were enacted in the 1980s. Tiger farms no longer supply a demand for body parts from makers of traditional Chinese pharmaceuticals. Responsible members of that industry want their ancient medicinal approach to gain an international market, which would be tarnished by the use of tiger parts; indeed, a domestic poll that Mills cites shows that most Chinese do not approve of tiger farming or its products, rightly fearing that this harms their nation's global image. Instead, the market is now based on luxury items like pelts, meat, sexual parts, and, not least, tiger bone wine (figure 8.6). Mills observed on one trip in 1994 that the number of tigers on farms was increasing, and later she estimated that by 2015 the number was between 5,000 and 6,000, with some farms having active wineries. Easily produced by soaking entire skeletons in cheap rice alcohol, tiger bone wine is now the most popular tiger consumer item. This has led to a highly lucrative criminal business in several Southeast Asian countries and that, in the case of China, is expressly sanctioned by an official arm of the government. As tigers have increasingly become rarer and worldwide pressure to protect them grows, traffickers in wild feline body parts have now begun to turn to common leopards, jaguars, and snow leopards to supply products derived from them as substitutes.

As of this writing, conservationists have received tragic news: the Chinese government, caving in to the greed of special-interest groups, has now legalized the use of tiger and rhino body parts for use in medicine as well as purported "research," nullifying the earlier prohibition on importing these. The already swollen black market for such products, aggravated

FIGURE 8.6. Tiger bone wine.

Falsely claimed to restore virility and cure diseases, products made from tiger and other wild-cat body parts are commonly found in markets and retail shops throughout several East Asian countries, including these wines made by soaking skeletons in alcohol. *Source*: Photo of skull courtesy of Dreamstime; photos of wine bottles by Li Quan.

by the state-sanctioned activities of tiger farms, will now make protection for wild tigers and other big cats, as well as for other endangered species, even harder. As John Goodrich, chief scientist at Panthera and senior director of the Tiger Program, said, "In reversing this ban, China has helped to legalize the execution and extinction of the magnificent tiger" (Environmental Investigation Agency 2019).

The exploitation of tigers in Asia is paralleled, though in a different way, by that of lions in some African countries like Tanzania and South Africa. Here state-licensed "lion farms," sometimes purporting to further tourism and education, are, intentionally or not, breeding facilities that cater to the desire of wealthy foreigners to pay to kill a trophy-quality, almost invariably male lion. On such farms the adult lions, like tigers in some Asian countries, are no more than commodities, like cattle or pigs. Although raised with adequate food, water, and sanitation, males are ultimately bred to be "hunted" at the age of four or five, sometimes lured by a bait of meat, to be shot in a fenced area of varying size, a practice sometimes known as "canned hunting." The "hunters" usually pay about US$35,000 for a successful kill. On some South African farms, females and males of unacceptable trophy quality are also sold by the *leeu boer* (Afrikaans for "lion farmer") to the lion-bone trade, where, like tigers, their bones and other body parts are used for reputed medical purposes. On a typical lion farm are scores of males that await hunting age—the adult females are simply breeders, with surplus cubs sold to petting zoos.

At some private wild animal parks, male lions, although not expressly raised for the canned hunting trade, may still end up as trophies as a result of problems related to the basic behavior of lions, which the captive-animal industry is mostly unable to deal with. In the wild, the resident pride male drives off the maturing two- to three-year-old males, which, as adults, might otherwise compete for dominance and mate with the pride's females. Such males become wanderers, sometimes forming coalitions with siblings or nonrelated males. These eventually try to usurp neighboring pride territories and, after defeating the established males, mate with the female lions—often (although not always) killing the cubs to bring the pride females into estrus. Thus, genetic variability is maintained. On a "managed" lion reserve with limited space, however, unless artificially removed these young males usually remain nearby, even if driven off by the resident male. In time they will also attempt a takeover when the vigor of the dominant male declines, with the result that they will be mating with their home pride's own aunts and mothers. The limited space of a managed reserve may create a different problem when the original, older male continues to survive without challenge, compounding inbreeding by mating with his own daughters. Since other reserves seldom want surplus lions, the reserve owner has limited choices: expensive contraception; euthanasia; creating a secondary, tourist-oriented cub petting zoo; or selling the surplus cubs to a *leeu boer* for canned hunting. No matter how large the private reserve, there is only a finite amount of space to sustain free-ranging predators like lions, so even if a reserve manager is opposed in principle to lion hunting he or she must deal with the excess lions that are produced under these conditions. With few exceptions, the captive-bred industry negatively affects wild lion populations since these individuals are often caught and smuggled across porous borders from source countries like Botswana to allay the decline in gene diversity of captive populations. The argument that captive lion breeding, as with that of tiger farms, takes pressure off wild populations is false and is contributing to the cats' decline.

Although they are not pantherins, Asiatic cheetahs (*Acinonyx jubatus venaticus*) are unfortunately, like some of their larger cousins, now critically endangered and for this reason are discussed here. Once, like the Asiatic lion, occupying a huge range, Asiatic cheetahs could be found from the Arabian Peninsula and the Near East across the Caspian region to India. The subspecies is now extinct in all of the former areas except in southwest Asia, mainly in Iran. In the past severe winters, uncontrolled hunting and conversion of natural grasslands into agricultural areas gradually reduced the overall population to about 200 in the 1970s, and now it is estimated that only about 50 individuals remain in three sub-populations , scattered over 140,000 sq. km. in the central plateau of Iran. As in the past hunting is a main threat, often from armed drug smugglers who consider cheetah skins and parts an additional source of income, while conflict with pastoralists, road construction companies and mining operations near protected areas disrupts the cats' hunting and breeding. Some hope currently exists through Indian government efforts (beginning in 2014) to reintroduce Iranian cheetahs into the western and northern parts of India, but as of this writing these suffer from Indian bureaucratic delay and inefficiency. The divided Iranian government has on one hand supported cheetah (and leopard) conservation, but on the other has jailed its own conservationists, and those from other nations, for the sake of partisan politics.

Natural ecosystems and common human land-use practices in or near protected areas are often incompatible, especially slash-and-burn, or fire-fallow, cultivation, in forest areas. It involves assarting—the chopping down and subsequent burning of natural woodlands to create a field called a swidden. This makes room for growing nutrient-hungry crops like corn, and the method has become standard in the rural regions of many tropical countries (figure 8.7). In historic cultures, such as that of the ancient Maya in Central America, a shifting pattern of cultivation, milpa, was instead the custom, in which small areas for agriculture were sporadically cleared and the plots used in sequence, anticipating the modern form of crop rotation. Such areas were left for adjacent areas of native forest to regenerate through natural plant and animal seed dispersal. This, and

FIGURE 8.7. Slash-and-burn (assarting) in Indonesia.
This type of agriculture, widely practiced in rural areas of some developing countries, is a fast, convenient way to clear land for growing and yields a successful harvest for the first season but after repeated growing seasons quickly exhausts the already nutrient-poor tropical soil. More and more plots must then be assarted to sustain yields, outpacing the surrounding forest's ability to regenerate the ravaged land. *Source:* Photo by Ethan Crowley.

the fact that ancient rural populations were smaller and far less concentrated than today's, made human habitation here sustainable, and such regions remained heavily forested. When cultivated on a continual basis in the same growing space, however, the soil quickly becomes exhausted and the plot must then eventually be abandoned for others where the process is repeated. In tropical biomes nutrients are located mostly within the vegetation rather than the underlying soils, which are relatively thinner and less fertile than those in temperate systems. Tropical soils are built up only from the constant, nonseasonal fall and recycling of dead leaves and decaying detritus from the living canopy over a period of thousands of years, and a given tract requires the close proximity of the forest to both replenish and recolonize the area to the point where it can once again thrive. Slash-and-burn agriculture has become common in many poor agricultural areas, and it could take decades, even centuries, for natural vegetation to come back, especially when the cultivated clearings are large. If these clearings are too greatly separated from nearby forest, it could take millennia. This unsustainable form of agriculture, along with illegal woodcutting for fuel from charcoal, is also gradually eating away wildlife habitats in tropical forests and temperate woodlands.

Attendant on this are the attempts by some palm oil growers, cattle ranchers, and other large corporations to lease and "develop" or "utilize" large tracts of virgin rain forest, occasionally facilitated by bribing state or national governments. This can create a spiral of destruction: in Brazil the Amazon rain forests are now being cut for timber, cattle grazing, and, increasingly, for soybean production as countries such as China, currently unable to purchase soybeans at their customary price from the United States because of a trade war, have turned to Brazilian soybeans. Aspiring politicians in some governments, wishing to garner popularity, may also promote the construction of seemingly beneficial but ill-conceived, eco-logically destructive dams and hydroelectric projects. In these schemes the flooding of an area to create a reservoir is misleadingly labeled as "creating more habitat" because of the resulting lake. In reality, damming in tropical lowlands destroys wildlife areas by forever flooding rivers and streams that serve as vital riparian habitat that supports large mammals like elephant and tapir, as well as prey-rich hunting territories for big cats like tiger and jaguar. In spite of their legal designations and official management poli-cies, many reserves and parks are vulnerable to such short-sighted political and corporate schemes. These are in effect paper parks, areas that the-oretically are protected but, through lax management or the time-hon-ored system of *baksheesh*, or kickbacks, suffer from the pillaging of timber and wildlife for profit and therefore are little more than lines drawn on a map. This condition is encountered most frequently in wildlife areas des-ignated "reserves" (which theoretically allow limited use by local inhabits for grazing and other sustenance purposes in some parts of the reserve) but can and still go on in a "park," which is supposed to be immune from exploitation.

A second serious form of human impact comes from expanding domes-tic animal husbandry in and bordering on designated wildlife reserves. In historic times, when natural and unspoiled areas were still huge, the small numbers of domestic ungulates were not a problem, but today swollen herds of cattle, sheep, and goats ravage the limited vegetation and water resources in many areas of Africa, India, and Central Asia, taking already scarce food from the small populations of native ungulates that may starve or die of thirst as a result. Thousands of sharp hooves cross and recross overworked, eroded, and aridified land, churning it into dustbowls in sum-mer and quagmires during the monsoon seasons (figure 8.8). In pastoral societies the number of animals owned is not just a source of income and a buffer against economic loss but is also is a sign of wealth and status, and local herders, in search of additional grazing or browsing, may often ille-gally bring their animals into protected areas.

When the first large reserves and parks were established by the Euro-pean governments early in the twentieth century in what were then still their colonies in Africa, management tended to be rigorous, and the solution to the problem of traditional but destructive land use was simply to move the ethnic peoples out. Pastoral societies such as the Masai, for example, some of whom lived within the Serengeti National Park's perim-eters, had long kept (and today elsewhere still maintain) the huge num-bers of cattle that can wreak such devastation on the land. Park authorities realized the controversial but inevitable need to remove and relocate

FIGURE 8.8. Herd of domestic cattle in India.
The combined grazing and browsing habits of domestic cattle, sheep and goats, concentrated by humans in numbers far exceeding any that would be found in nature, over the years destroy the natural vegetation that provides food for native ungulates. The many hooves in constant movement churn up and degrade the surface of the soil, making it difficult for native plants to regenerate.
Source: Photo courtesy of FeaturePics.

the settlements. As a result, in 1959 entire villages were resettled in the nearby Ngorongoro Conservation Area and, while there was sometimes deceit on the part of the authorities, the Masai in most cases were fairly compensated with equivalent amounts of land and monetary payments or goods. Although the long-term goal of protecting wildlife was justified, the Masai, who had lived on the Serengeti for some 200 years, were seldom if ever given an opportunity to participate in the issues of relocation and saw this as the takeover of a significant natural resource by the authorities for the purpose of allowing only rich, privileged foreign visitors to use the park. Although the Masai had historic relationships with the land and native wildlife, their pastoral practices over time were becoming unsustainable within a huge but nevertheless fragile natural ecosystem. In Belize, Alan Rabinowitz was faced with the same issue when he neared the end of his study that led to the Cockscomb Basin's designation as a national forest reserve. Although during his research he had become an intimate neighbor of and grown to love the local of the Maya peoples, who had occupied this part of Central America for thousands of years, he realized that their current slash-and-burn agricultural practices were slowly but surely destroying what was perhaps the most jaguar-dense forest region of South America. In light of this, Rabinowitz finally and reluctantly advocated for the negotiated compensation and resettlement of the region's inhabitants, and, following this, the Cockscomb Basin Widllife Sanctuary and Jaguar Preserve was designated in 1984.

The confinement of wild animals to reserves and parks, although intended to offer them protection, is only a stopgap measure in their ultimate conservation and, in the long term, is as unsatisfactory as the human abuses that led to the need for their creation. With the possible exception of places like the Serengeti and Arctic National Wildlife Refuge, few such designated areas are vast enough to encompass an entire ecosystem for its resident creatures. Animals like big cats usually need huge home ranges to satisfy their requirements, and when young animals disperse to find mates and new hunting territories, they are apt to venture outside park boundaries and come into conflict with human interests. As human populations and the land requirements needed to feed them soar in those regions where big cats still survive, first farms and ranches then towns and cities may one day ultimately encircle the protected areas, reversing the conditions that existed as late as a few hundred years ago in which human settlements were mere punctuations in nature's vast meadows. It's a strange, almost surreal sight to see lions in Kenya's Nairobi National Park silhouetted against the tall buildings of downtown Nairobi less than six miles away (figure 8.9). Other nations have similarly shown great farsightedness in conserving their natural environments: China, in spite of its disastrous policy regarding the tiger body-parts trade, has given considerable protection to the Tibetan Plateau, and in Xinjiang, Qinghai, and Tibet there are now 700,000 km^2 of protected land.

Whether or not one agrees with the taxonomic bases of their trinomial labels, all big cat subspecies share strong, closely linked genetic similarities, and this has given tigers, lions, leopards, and jaguars potential resilience in adapting to changing conditions over geologic and historic time. However, if pantherin populations are confined to a reserve or park and geographically isolated from other populations, they also become genetically isolated and vulnerable to any disease or harmful conditions that may occur. This is what befell steppe lions and some Southeast Asian subspecies of tigers (see chapter 7), and, as with today's cheetahs, the resulting genetic bottlenecks made them vulnerable to extinction by limiting their ability to adapt. Thousands more lions now exist in isolated populations on private game sanctuaries in U.S. states like Texas than on the African savannas, and most of the world's living tigers are captives. Although they may be well enough cared for, it is debatable whether these animals are still truly lions or tigers or merely caged sideshows. As we have seen throughout this

FIGURE 8.9. In Nairobi National Park, Kenya, among the smallest of African protected wildlife areas, a lion gazes toward the looming city skyline.

book, an animal is the unique outcome of an almost immeasurably long evolutionary experience. If a captive big cat can no longer hunt, disperse, or interact with its prey or others of its kind, has it not become in a sense yet another domestic animal? Biometric measurements of the jaws and teeth in past studies of some zoo-bred lions show decreases in bone density and tooth alignment, suggesting that degenerative morphological changes over generations follow the abandonment of an active lifestyle for a sedentary one. In humans, the skeletal density and bioapatite concentrations of Neanderthal bones are typically much greater than in the equivalent bones of our own species; this is interpreted as the Neanderthal's need for greater strength in the intensely physical, close-quarter hunting tactics used for killing prey. Any wild animal species maintains its integrity by virtue of interacting with the totality of its environment, and it is possible that we might not wish to see the progeny of zoo-bred tigers, lions, or leopards in the centuries after their extinction in the wild (figure 8.10).

In light of this, what can be done to save pantherins now and within the foreseeable future? The conservation of big cats and their worlds can be symbolized by Durga, the namesake of the imaginary tiger of Sikhote-Alin Reserve and the fierce warrior and mother goddess of the Hindu pantheon, also known as Devi or Shakti or by other names. Her many arms carry weapons to defeat the forces of greed and destruction that threaten nature and the cosmic good, and she is depicted as a woman without fear, riding a

FIGURE 8.10. Degenerate tiger in future zoo.

tiger or lion (plate 12). Inspired by Durga's passion and creative strength, we suggest that the time has come for a spirit of wild conservation in which the worlds of legitimate human need and nature are not seen as separate, competing entities but instead will coexist. In this scenario, sustainable land stewardship and environmental responsibility are key concepts. Those who claim the right to live on or near a natural area must be helped and encouraged to do so without eliminating its native animals and plants. Making this possible will demand, in addition to more-effective and better-funded antipoaching efforts, far-reaching changes in both official government policies and longstanding land uses in wild areas. The latter is hardest for ethnic, rural societies that have known no other agricultural or pastoral practices than the traditional ones they were raised with, and it requires both active assistance in adopting alternatives and extensive financial aid from affluent Western nations (figure 8.11).

FIGURE 8.11. Conservation of snow leopards.
One of several organizations working to protect big cats, the Snow Leopard Conservancy has been successful in helping local villages in Nepal and India improve their community incomes and living standards in partnership with programs that further the welfare of native animals and habitats. Here Rodney Jackson adjusts a sedated cat's radio transmission collar. *Source*: Photo and logo courtesy of Rodney Jackson and the Snow Leopard Conservancy.

As Schaller has pointed out in his 1998 book *Wildlife of the Tibetan Steppe*, conservation problems are mainly social and economic, not scientific. The needs of pastoral societies in particular are not easy to solve, and the practice of keeping large numbers of cattle and other domestic animals sometimes has both deep economic roots and cultural ones that require diplomacy and innovation when dealing with long-term attitudes. In such areas as Tibet's Chang Tang Reserve, livestock that sustains a family may, in addition, bring in vital income from wool, butter, and milk and is an essential buffer against economic misfortune. Killing domestic animals for food is rare in most pastoral cultures because it not only cuts into the number of head needed to produce renewable food for sale and sustenance but also undercuts the potential for herd recovery if adult or juvenile mortality is severe. Usually a premium is placed on maximizing the numbers of animals in a herd. The unfortunate result is that human population expansion results in overgrazing of traditional pastures, which in turn leads to dispersal into areas that can only marginally support domestic animals, driving native ones from their food resources.

In some pastoral societies, such as those of East Africa, livestock is a source of not only sustenance and profit but also prestige. In this case, efforts can be made to educate settlements in what the land can sustain, and the government can work to convince them of the economic benefits and other advantages of wildlife tourism. At the same time, the damage done by hungry cattle, goats, and sheep can be limited both by providing the herders with incentives to curtail livestock numbers and by compensating them for occasional losses from big cats. Retribution killing following loss of livestock to feline predation could be averted if owners kept guard dogs, prevented their animals from wandering into prohibited areas, employed better herding techniques, and confined their livestock in guarded corrals at night. Rabinowitz reported that the villages that adopted these practices usually lost few animals to jaguars during his time in Cockscomb. Extensive fencing, which is a costly and difficult adjustment for herders whose animals have traditionally been allowed to forage across large areas, is undesirable because it also restricts natural wildlife migrations. The misguided idea of building a wall to restrict illegal immigration along the U.S.-Mexican border would, for example, put an end to jaguar repopulation of the American Southwest.

In the case of tropical and subtropical regions, educating rural communities to abandon slash-and-burn techniques for a historically sustainable milpa system of crop rotation is a first step in stopping the soil exhaustion, erosion, and destruction of forests. Along with this comes the need for alternatives to using trees for fuel in the form of simple devices like solar ovens; other new technologies can aid in providing clean drinking water and composting human waste. Private organizations such as Oxfam, Heifer International, and World Neighbors have sprung up to meet these and other needs, focusing on training and education in communities as well as donating necessary items. In the case of Heifer International, both donations and direct assistance have helped Peruvian villages to return to raising guinea pigs as a protein source instead of snaring and capturing agoutis and other potential jaguar prey. Threats to tropical and subtropical ecosystems come from without as well as within, however. In Southeast Asian, Central African, and South American countries, large timber, palm oil, cocoa, and other giants of the extraction and growing industries remain poised to lobby national governments to allow them to raze and lease forest land for commercial gain. Deforestation contributed to the extermination of Balinese and Javan tigers and continues to loom over rain forest sanctuaries in the rest of Southeast Asia and South America. Wholesale tropical forest destruction has recently been successfully fought by environmental advocacy groups like the Wildlife Conservation Society (WCS), the World Wildlife Fund (WWF), Rainforest Action Network (RAN), the Rainforest Alliance, and, most recently, Azuero Earth Project. These and others seek to educate Western buyers to boycott products containing unsustainably harvested ingredients while, at the same time, promoting small-scale, forest-compatible growing schemes that are owned by and benefit rural villages. Working with cattle ranchers in the tropical dry forests of Los Santos, Panama, Azuero Earth Project shows owners which native trees to plant on the land in order to form a corridor that connects with established forest and restores forest habitat for wildlife. With the public's active financial and consumer support, such advocacy groups can help to replace illegal, forest-destructive harvests of coca, opium, and other potentially harmful substances with healthy crops like sustainably raised coffee and cocoa that provide income for the human inhabitants and do not destroy forests.

Schaller and other conservation biologists have suggested that the goals of protecting wildlife and accommodating traditional human land use can sometimes be reconciled by adopting more flexible attitudes toward land use. This means that in such reserves as Chang Tang, Tibet, certain core areas would be restricted against *any* human incursion (other than for management and scientific study) because the type of habitat is so sensitive and because the needs of wildlife, such as undisturbed seasonal breeding, are so critical. In some less ecologically sensitive peripheral areas, however, practices such as limited pastoralism and other small-scale activities for the direct benefit of the local residents could take place. The main goal would be to help herders increase their economic returns from livestock, thereby offsetting restrictions on the number of animals kept. Initiatives could include a price-support system that would allow households to sell wool, dairy, and other commodities at higher prices directly to buyers and the establishment of international markets for the sale of crafted items. Because it can disastrously restrict the free movement of wildlife, any extensive fencing would be prohibited in a reserve or park in favor of management policies that adopt guard animals and other alternatives for domestic livestock protection, and locals would be compensated by the government for animal loss from occasional predation.

Other uses would be more problematic and would need close supervision to avoid abuse. Currently depleted local wild ungulate populations in some peripheral reserves of national parks *might* at some point recover enough to allow strictly controlled hunting for sustenance in the future, thereby providing a needed food source; limited commercial trophy hunting, on a likewise strictly regulated basis and monitored by outside agencies such as CITES, could help offset the typical underfunding of reserves in developing countries. Harvestable wildlife could be managed jointly by the local community and the government, with regulations and quotas determined on a local, year-to-year basis (this would not apply to big cats, however, for reasons described earlier in the chapter).

In the end, the success or failure of such a multiuse wildlife sanctuary will depend on winning over the inhabitants who live near such reserves and on instilling within them the idea that such areas are a part of *their* heritage and culturally and spiritually a direct part of *them*. The days are long gone when a reserve could be summarily isolated for the enjoyment of elite, outside visitors while placed "off-limits" to the local inhabitants who had a historic relationship to the land and its wildlife. Each reserve or park will inevitably pose its own unique problems and challenges, and education programs for the neighboring people will be needed in each case to bring home the practical economic benefits of good land stewardship and wildlife management.

Even more far-reaching than the establishment of regional core and peripheral sanctuaries for big cats and other species is the concept of creating "wildlife corridors," originally proposed by the conservation biologist Michael Soulé (University of California, Santa Cruz) and supported by Schaller and other workers. Soulé and his colleague Reed Noss have pointed out that by steadily restructuring most of the planet's biosphere, humans have effectively stopped the process of natural evolution by preventing the dispersal and migration of large animals, including predators. Big cats, wolves, and bears are keystone species that help maintain natural ecosystems through their direct effect on herbivores and indirect effect on plant communities; their presence is vital to the health of these areas. In the vision of Soulé and other proponents, wildlife corridors of natural, undeveloped land would connect key regions of biodiversity, sometimes already existing as parks and reserves, with one another (figure 8.12). Where they overlap national boundaries, these corridors would be cooperatively protected and managed by neighboring countries to allow for the dispersal and migration of animals. This would help to once again reestablish formerly continuous but now fragmented big cat populations, extending their ability to disperse and find new hunting territories and breeding partners. Soulé is a founder of the Wildlands Network, an

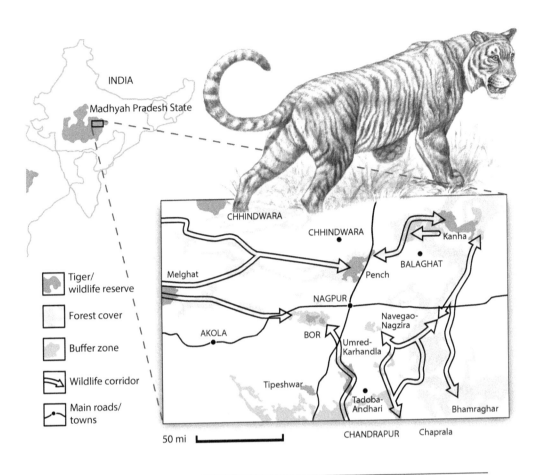

FIGURE 8.12. Proposed corridors connecting tiger populations in Madhya Pradesh state, India.
Source: Compiled from Google-sourced online maps, courtesy of Panthera.

organization that promotes the idea of enabling wildlife to roam across a continent. Examples of such conservation goals are "Y2Y," a vast swath that would interconnect Yellowstone National Park in the United States to Yukon Province in Canada, and the "Peace Park," a project on which Schaller has been patiently working since 2007 with Pakistan, Afghanistan, Tajikistan, and China, that would establish the planning and management of an incredible 32,000 km² (20,000 sq. mi.) of extended corridor across the heart of Asia. If the participating nations can work effectively together, this will protect such iconic creatures as the legendary ghost of the Palmirs and Himalayas, the snow leopard, as well as Marco Polo sheep and other regionally threatened species. On a smaller scale is the proposed corridor that would allow pumas, presently confined to the Santa Monica Mountains of Los Angeles, to disperse northward into the

huge, adjoining Los Padres National Forest. The pumas, although flour-
ishing, have now reached a saturation point in the numbers that can opti-
mally hunt and live in the Santa Monicas, and the project would let them
freely move across a partly elevated pathway of reforested landfill to hunt
for mule deer and find mates. Projects like these and others in the pantherin ranges of Africa and South America would, if successful, alleviate
the problem of genetic bottlenecks.

Other visions include someday bringing back the historic tiger, lion,
and leopard subspecies now lost to human-caused extinction. Break-
throughs in the understanding of genetics have led researchers to discover
ways of modifying—in effect editing—an organism's original genome, the
total and distinctive "recipe" coded within DNA that determines the way a
species as well as an individual looks and behaves. One of these techniques
is based on the existence of a structural DNA feature called clustered
regularly interspaced short palindromic repeats, mercifully shortened to
CRISPR. This refers to the protein-based sequences within a gene's strand
of DNA, and the simplest and most commonly used lab technique to now
alter a genome is CRISPR/Cas9. In this biochemical process one or more
nodes located along a section of the DNA sequence that specifically determines an organism's particular character is "snipped" away or removed by
interjecting modified enzymes, combined with a "guide" or synthetic form
of RNA (the "messenger" twin of DNA), on its exact location in a particular
gene. The section containing the original node or node sequence is then
replaced with a modified section that is responsible for a different character. The artificially reengineered gene then becomes a permanent part
of the organism's genome and can then be passed on, through selected
artificial breeding, to other individuals with similarly modified genes to
produce offspring with this new characteristic. For example, the genes that
permit an *Anopheles* mosquito to host a harmful malaria plasmodium or
Zika virus in its blood could be altered and replaced. The genetically modified mosquitoes would then be selectively bred in the lab and released into
the wild in huge numbers. The mosquitoes with the artificially selected
character would reproductively overwhelm the unaltered ones that carry
malaria or Zika and, within a short time, disease-free insects would totally
replace the disease carriers.

For tigers, this process could take genetically modified embryos bearing such characters as the long facial ruffs and belly fringe of the Caspian
tiger and, through in vitro implantation into a female from another, extant
subspecies, produce individuals morphologically resembling the extinct
subspecies. Cubs might someday be successfully raised to once again live
and reproduce in their former historic ranges, provided that they could be
trained to hunt and survive in the wild. A carefully applied CRISPR program could be used as a shortcut to the slow process of selectively breeding individuals of certain zoo-bred lion populations that appear to have
the physical characters of extinct subspecies like Cape and Barbary lions.

FIGURE 8.13. Reconstructed Caspian tiger in its former habitat.
In a scene that may one day become a reality, two reconstructed Caspian tigers, genetically altered and bred from Amur (Siberian) tigers, relax along the reed-covered banks of the Ili River, Kazakhstan. This and some other nearby areas of their once huge natural range in western Asia were among the Caspian tigers' last strongholds and are a part of the Kazakhstan government's long-range plans to reintroduce this magnificent cat to where it once roamed.

The recent discovery that the genome of the endangered but still extant Amur (Siberian) tiger has great similarity to that of the extinct Caspian *Panthera tigris virgata* allows for hopeful speculation that CRISPR could help bring back this magnificent animal, which could then be reintroduced into parts of its historic range that are still in a natural state (figure 8.13). Could we again see these cats prowling the marshes of western Asia? And might the Indochinese islands once more have Balinese and Javan tigers? The drawback to this exciting possibility is that the animals would require the existence of suitable habitats large enough to sustain breeding populations, a condition perhaps no longer feasible in crowded Indonesia. Only if adequate natural habitat could be provided would it be practical and ethical to reestablish a genetically reconstructed subspecies.

Also, the complexity of feline behavior forces a question: Even if an animal looks morphologically correct, what intrinsic and perhaps forever unknowable learned behaviors and instincts were lost with the death of the last of these cats? One only has to consider the difficulty that Joy Adamson and her husband George had in training orphaned lions for reintroduction to the wild in Tsavo, Kenya, during the 1950s (Adamson 1960). These had to be helped to hunt for themselves and to be taught to successfully inter-act with others of their kind. Could genetically resurrected animals survive in a meaningful way, or would they only be living stage props, brought back merely to satisfy a human longing for what can no longer be? We must consider the specter, mentioned earlier in the chapter, of a zoo-bred cat once again living in the wild but no longer its true self.

Equally exciting but even more fraught with problems is the possibility that genetic engineering could someday bring back an extinct cat from the ice ages. Unlike the somatic cell nuclear transfer process (popularly known as cloning) that now routinely permits the individual duplication of pets and other mammalian breeding stock, the ancient DNA of even the most well preserved of prehistoric cadavers is usually contaminated, and its integrity is compromised by DNA from outside sources. It is invari-ably too degraded to allow it to be used in its original state. Many genes that determine physical appearance, physiology, and behaviors are apt to be damaged or missing. If any key characters *have* been preserved in the extinct animal's bone or other tissues, they *may* be amplified and reinte-grated into the equivalent genetic sequence of a related species through the CRISPER/Cas9 process. For example, key characters of the Caspian tiger could be introduced into the genetic sequence of an Amur tiger. Unlike the reasonably well known characters of a Caspian tiger, however, the reconstructed genome of a steppe lion would still have gaps in its gene sequences, the restoration of which could only be approximated during genetic engineering. If cellular duplication *could* be achieved, there might still be several technical hurdles in ultimately producing a live, healthy steppe lion cub, as explained by Beth Shapiro (University of California, Santa Cruz) in relation to the goal of recreating a living woolly mammoth in her 2015 book, *How to Clone a Mammoth*. In the case of a steppe lion, the need for a surrogate female of a closely related species in whose uterus the embryo could be implanted might seemingly be easily solved by an implantation into a savanna lion. With evolutionary divergence, however, barriers can arise, and one is that chemical differences in the surrogate's uterus and placenta may not allow the embryo to develop; here, genetic editing may cause problems by interfering with the reactions between the embryo and the maternal host. Even before implantation, if some factor in the reprogrammed nucleus of the reedited extinct species is incompatible with the egg cell into which it's injected (such as a mismatch between the nuclear and mitochondrial DNA), cell division will not take place, and no embryo will develop.

If we do somehow actually engineer a living steppe lion from preserved tissue, the genome, although perhaps mostly representative of the species, will only be that of a single individual, and subsequent clones from this would have exactly the same genetic makeup. Unless the entire scope of steppe lion genetic variability is known by then from other preserved indi-viduals to allow CRISPR tweaking, there would be no point in creating a "population" of carbon-copy cats whose natural behaviors would be unknowable.

Although resurrected steppe and American lions may be an impossible dream, modern African and Asian lions have factored into an equally daring but more feasible possibility known as Pleistocene rewilding, proposed as early as 1967 by such wildlife workers and paleobiologists as the late Paul S. Martin and, later, Tim Flannery (Macquarie University). This concept is an extension of the already current practice of reintroducing locally extinct species back into areas they once inhabited in historic times and has been successfully accomplished with muskoxen, which until recently were relegated to the Canadian Arctic but have now been relocated back to Finland and Russia, where they formerly lived. Martin, Flannery, and others note that the present biological communities of North America are ecologically unbalanced and do not function in a natural way because they lack large mammals. This extinction occurred at the end of the last glacial and left the fauna in a state of zoological impoverishment between 12 and 10 ka. Pleistocene rewilding would seek to correct this imbalance by introducing related but nonnative stand-ins for extinct species that approximate the latter's ecological roles as closely as possible. Examples among herbivores would be the importation of extant dromedary camels (*Camelus dromedarius*) into the American Southwest to replace the now extinct camelid (*Camelops hesternus*), which browsed sage, the salt bush *Atriplex*, and other lower-elevation desert plants; llamas like the guanaco (*Lama guanicoe*) or the vicugna (*Vicugna vicugna*) could replace the stout-legged llama (*Palaeolama mirifica*) in mountain habitats. The mixed browse- and grass-feeding Asian elephants (*Elephas maximus*) might be good candidates as fill-ins for their extinct grazing relatives the Columbian mammoth (*Mammuthus columbi*), whose feeding habits conceivably maintained the ecological balance between woodlands and prairies and helped to disperse the seeds of native Osage orange (*Maclura pomifera*) and other native plants when these were passed through the digestive tract. In the case of carnivorans, modern cheetahs could again fill the niche of extinct American cheetahs (*Miracinonyx*), which coevolved with and were the only predators able to catch the pronghorn antelope (*Antilocapra americana*), which can outrun any living animal pursuer and now has no native predators other than humans. The challenge, of course, would be to ensure that the introduced cheetahs didn't develop a taste for defenseless, slow-moving domestic livestock.

Even more controversial, however, is the idea of bringing in existing large feline pantherins to act as a check on expanding herds of such ungulates as bison, deer, and feral pigs that presently lack effective predation. Some U.S. residents, already opposed to having grizzly bears, wolves and naturally dispersing native pumas as neighbors in previously uninhabited "suburban-wild interface" areas in California, might be horrified by the proposal to introduce African or Asian lions into the Midwestern prairies, unaware that American lions were once natural predators on bison herds as late as 11,300 years ago. In some "developing" nations big cats and other carnivorans have been a historic hazard to which the culture adapted, but Western societies have generally lacked contact with large, potentially dangerous animals for hundreds of years. As a result North Americans and Europeans might be challenged by the idea of being near lions or jaguars when they camped in a national park or wildlife reserve in their home country, although they would happily stay in tented camps on the Serengeti. But is this so different from traveling to Yellowstone or Denali National Park to see Alaskan brown bears and wolves in their natural state? Alaska and certain parts of Montana, Idaho, and Wyoming are perhaps the only regions of the United States where Americans not only accept the presence of large predators but thrill at the chance to see them living as natural and free creatures (figure 8.14).

As unlikely as the idea of Pleistocene rewilding might seem, it brings us full circle to the dream symbolized by the god Durga and her pantherin steeds: sensitivity and compassion toward all living things is a universal, human value that transcends any one culture and goes beyond our own moment. Can we live with and accept the presence of big cats and in fact

FIGURE 8.14. Reintroduced lions in an American national park.
Could Asian or African lions one day take their place in a Canadian or U.S. national park like Yellowstone or Grand Teton as the ecological equivalents of the extinct American lion of 12,000 thousand years ago? As with grizzly bears and gray wolves, it can be argued that such predators have a natural place in such ecosystems.

FIGURE 8.15. Ancient pantherins in courtship.
Modern male and female pantherins like tigers can show great tenderness toward each other at times, often during their courtship prior to mating. Was it the same for vanished forms like the ancestral tiger *Panthera zdanskyi* and others? We will never know, but if we save modern species these will give us the chance to learn about such behavior.

all of wild nature without the urge to diminish, exploit, and destroy? As we witness the ever-accelerating elimination of many of the Earth's natural species and their habitats, we must remind ourselves that in great part it is we, as citizens of the planet, who must begin to reverse this loss. We can personally create meaningful and lasting change through intellectual and spiritual awareness; election of intelligent, informed government representatives; and support of environmental activism. Adjusting the everyday choices we make as consumers can make this a reality. In considering the long, winding evolutionary path of the big cats, we should not only treasure the beauty of these magnificent creatures but also resolve that their journey through time will not now end because of us (figure 8.15). In working toward that outcome, a quote by George Schaller is perhaps the best way to conclude this book:

> *Remember, there are no victories in conservation. You may have a temporary "looks good." Suddenly things change and the fight begins all over. So we've got to continue fighting.*

APPENDIX 1
Distribution of Pantherins and Other Felids in Geologic Time

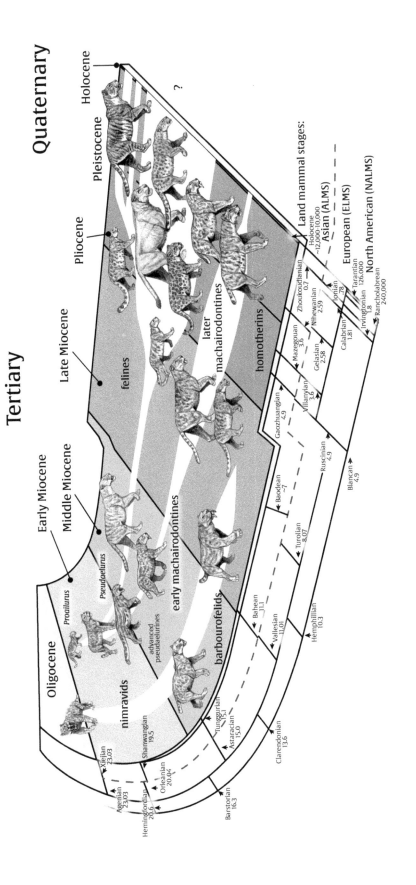

APPENDIX 1. Distribution of pantherins and other felids in geologic time.

APPENDIX 2
Pantherin Dispersals Across the World

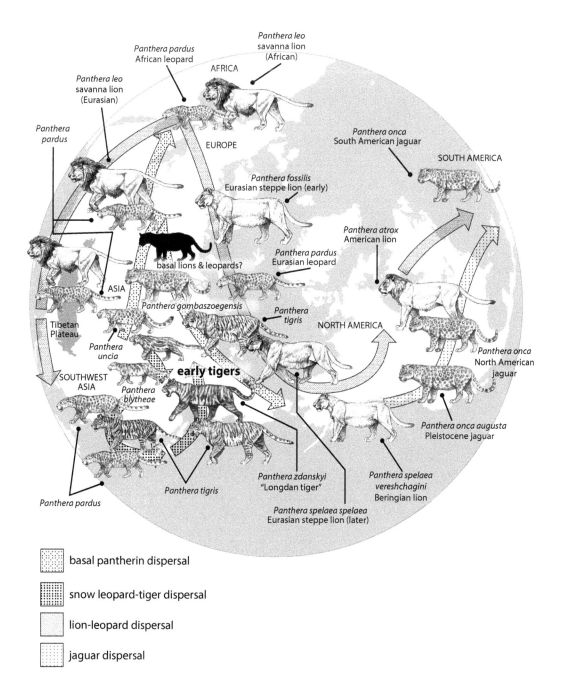

Panthera leo savanna lion (African)

Panthera pardus African leopard

AFRICA

Panthera leo savanna lion (Eurasian)

Panthera onca South American jaguar

SOUTH AMERICA

Panthera pardus

EUROPE

Panthera fossilis Eurasian steppe lion (early)

Panthera atrox American lion

Panthera pardus Eurasian leopard

basal lions & leopards?

ASIA

Tibetan Plateau

Panthera gombaszoegensis

Panthera tigris

NORTH AMERICA

Panthera uncia

Panthera onca North American jaguar

early tigers

SOUTHWEST ASIA

Panthera blytheae

Panthera onca augusta Pleistocene jaguar

Panthera tigris

Panthera zdanskyi "Longdan tiger"

Panthera spelaea vereshchagini Beringian lion

Panthera pardus

Panthera spelaea spelaea Eurasian steppe lion (later)

basal pantherin dispersal

snow leopard-tiger dispersal

lion-leopard dispersal

jaguar dispersal

APPENDIX 2. Pantherin dispersals.
Pantherins appear to have originated in Asia during the late Miocene. After the divergence of *Neofelis* (clouded and Sunda leopards) about 7 Ma, the center of pantherin radiation seems to have been the Tibetan Plateau. From this location the snow leopard–tiger lineage dispersed into northern, eastern and southeastern Asia while the jaguar-lion-leopard lineage migrated to Africa and then radiated out into Eurasia and the New World. As the machairodont sabertooths declined, pantherins became dominant in open woodlands and savannas, achieving an enormous geographic range that persisted until the megafaunal extinctions in Africa about 400 ka, in Europe about 30 ka, and in North America about 12 ka.

APPENDIX 3
Taking Action

If you are reading this, you are now among the last generation of humans that has any chance of ensuring the survival of Earth's tigers, lions, jaguars, common leopards, snow leopards, and many other wild predators. As these animals' habitats shrink, in part because of human population pressure and in part because of human ignorance, preoccupation, and greed, the end of the twenty-first century will bear witness to either the big cats' effective extinction in the wild or their continuance as natural, free-ranging creatures. Depending on your own resources, time, and willingness to be involved, as an individual you can actively help to determine this outcome through taking action by contributing toward or indirectly participating in the work of the conservation organizations aimed at saving big cats. Among the most effective are:

THE WILDLIFE CONSERVATION SOCIETY

2300 Southern Blvd.
Bronx, New York 10460
(718) 220-5100
https://www.wcs.org

RAINFOREST ACTION NETWORK

425 Bush Street
Ste. 300
San Francisco, CA 94108
(800) 989-7246
https://www.ran.org

PANTHERA USA

8 West 40th Street
18th floor
New York, NY 10018
(646) 786-0400
https://www.panthera.org/

THE INTERNATIONAL SNOW LEOPARD TRUST

4649 Sunnyside Ave N,
Suite 325
Seattle, WA 98103
(206) 632-2421
https://www.snowleopard.org/

THE EAST AFRICAN WILDLIFE SOCIETY

Riara Road, Off Ngong Road
PO Box 20110
00200 City Square
Nairobi, Kenya
+254 203874145
https://eawildlife.org/

WORLD WILDLIFE FUND

1250 24th Street NW
Washington, DC 20037
https://www.wwf.org/

APPENDIX 4
Species of the Genus *Panthera*

SCIENTIFIC NAME

Panthera atrox: late Pleistocene giant "American" lion.

Panthera blytheae: late Miocene/early Pliocene pantherin about the size of a modern lynx or serval; ancestor of the snow leopard.

Panthera fossilis: early-middle Pleistocene steppe lion or "cave lion," known from the early Pleistocene Kuznetsk Basin site of western Siberia.

Panthera gombazoegensis: early Pleistocene Eurasian jaguar-like pantherin.

Panthera leo: extant savanna lion.

Panthera leo azandica: extant Congo lion, from the northeastern Democratic Republic of Congo.

Panthera leo bleyenberghi: extant Angolan lion, from southern Democratic Republic of Congo and neighboring parts of Zambia and Angola.

Panthera leo kruegeri: extant Transvaal lion, from the Kalahari region to eastern South Africa.

Panthera leo leo: extinct Barbary or Atlas lion population from northern Africa.

Panthera leo melanochaita: extinct Cape lion, from South Africa.

Panthera leo nubicus: extant Nubian lion, from east and northeast Africa.

Panthera leo persica: extant Asiatic, Indian, or Persian lions.

Panthera leo senegalensis: extant Senegalese lion, from West Africa east to the Central African Republic.

Panthera onca: extant jaguars.

Panthera onca arizonensis: extinct Arizona jaguar.

Panthera onca hernandesii: extant western Mexican jaguar.

Panthera onca veraecrucis: extant southeastern Mexican jaguar.

Panthera palaeosinensis: the earliest basal jaguar-leopard-lion-like pantherin.

Panthera pardus: extant leopard.

Panthera pardus delacouri: extant Indo-Chinese leopard.

Panthera pardus fusca: extant Indian leopard.

Panthera pardus japonensis: extant north Chinese leopard.

Panthera pardus kotiya: extant Sri Lankan leopard.

Panthera pardus melas: Javan leopard

Panthera pardus nimr: extant Arabian leopard

Panthera pardus pardus: extant African leopard.

Panthera pardus saxicolor: extant Persian leopard.

Panthera pardus sindica: extinct Balochistan leopard.

Panthera pardus tulliana: extinct Anatolian leopard, once native to western Turkey.

Panthera spelaea: extinct Eurasian steppe lion, also known as a "cave lion."

Panthera spelaea intermedia: middle Pleistocene population of distinctly smaller-sized *P. spelaea* from the Igue des Rameaux locality in southwestern France.

Panthera spelaea spelaea: European or Eurasian steppe. (or cave lion) from a variety of Eurasian sites beginning at about 300 ka.

Panthera spelaea vereshchagini: the "East Siberian" or "Beringian" steppe lion, found in Yakutia (Russia), Alaska (United States), and Yukon Territory (Canada), from about 350 ka.

Panthera tigris: extant tiger.

Panthera tigris altaica: extant Amur (or Siberian) tiger.

Panthera tigris amoyensis: extant South China tiger.

Panthera tigris balica: extinct Balinese tiger.

Panthera tigris corbetti: extant Indo-Chinese tiger.

Panthera tigris jacksoni: extant Malayan tiger.

Panthera tigris soloensis: the extinct "Ngandong tiger," known from a Pleistocene locality near Ngandong, Indonesia.

Panthera tigris sondaica: extinct Javan tiger; according to some workers, encompasses Balinese and Sumatran populations and is one of only two evolutionary tiger groups with *P. t. tigris*.

Panthera tigris sumatrae: extant Sumatran tiger.

Panthera tigris tigris: extant Indian or Bengal tiger; according to some workers, one of only two evolutionary tiger groups with *P. t. sondaica*.

Panthera tigris trinilensis: extinct "Trinil tiger," known from the middle Pleistocene Trinil locality in Java, Indonesia.

Panthera tigris virgata: Extinct Caspian, Hyrcanian, or Turan tiger.

Panthera uncia: extant snow leopard.

Panthera youngi: fossil steppe lion found at some of the same levels as the hominin *Homo (Sinanthropus) erectus* ("Peking Man") at the Zhoukoudian (Choukoutien) Caves near Beijing in northeast China, from 700–200 ka.

Panthera zdanskyi: the earliest tiger, known from early Pleistocene fossils in northern China.

COMMON NAME

African leopard: *Panthera pardus pardus*
American lion: *Panthera atrox*
Amur leopard: *Panthera pardus orientalis*
Amur tiger: *Panthera tigris altaica*
Anatolian leopard: *Panthera pardus tulliana*
Angolan lion: *Panthera leo bleyenberghi*
Arabian leopard: *Panthera pardus nimr*
Arizona jaguar: *Panthera onca arizonensis*
Asiatic lion: *Panthera leo persica*
Atlas lion: *Panthera leo leo*
Balinese tiger: *Panthera tigris balica*
Balochistan leopard: *Panthera pardus sindica*
Barbary lion: *Panthera leo leo*
Bengal tiger: *Panthera tigris tigris*
Cape lion: *Panthera leo melanochaita*
Caspian tiger: *Panthera tigris virgata*
cave lion: *Panthera spelaea*
Congo lion: *Panthera leo azandica*
Eurasian or European steppe lion: *Panthera spelaea spelaea*
far eastern leopard: *Panthera pardus orientalis*
Hyrcanian tiger: *Panthera tigris virgata*
Indian leopard: *Panthera pardus fusca*
Indian lion: *Panthera leo persica*
Indian tiger: *Panthera tigris tigris*
Indo-Chinese leopard: *Panthera pardus delacouri*
Indo-Chinese tiger: *Panthera tigris corbetti*
Javan leopard: *Panthera pardus melas*
Javan tiger: *Panthera tigris sondaica*
Malayan tiger: *Panthera tigris jacksoni*
Ngandong tiger: *Panthera tigris soloensis*
north Chinese leopard: *Panthera pardus japonensis*
Nubian lion: *Panthera leo nubicus*
ounce: *Panthera uncia*
Persian leopard: *Panthera pardus saxicolor*
Persian lion: *Panthera leo persica*
savanna lion: *Panthera leo*
Senagalese lion: *Panthera leo senegalensis*
Siberian leopard: *Panthera pardus orientalis*
Siberian tiger: *Panthera tigris altaica*
snow leopard: *Panthera uncia*
south China tiger: *Panthera tigris amoyensis*

southeastern Mexican jaguar: *Panthera onca veraecrucis*
Sri Lankan leopard: *Panthera pardus kotiya*
steppe lion: *Panthera spelaea*
Sumatran tiger: *Panthera tigris sumatrae*
Transvaal lion: *Panthera leo kruegeri*
Trinil tiger: *Panthera tigris trinilensis*
Turan tiger: *Panthera tigris virgata*
western Mexican jaguar: *Panthera onca hernandesii*

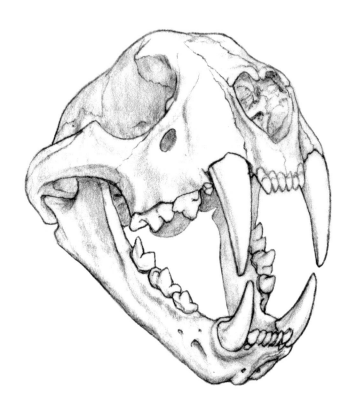

Skull of mainland clouded leopard, *Neofelis nebulosa*.

GLOSSARY

Acinonychini: tribe of big cats containing the puma, cheetah, and jagua-rondi (**acynonychins**).

Aeluroidea: mammalian superfamily containing cats, civets, mongooses, and hyenas (**aeluroideans**).

age: interval of geologic time that represents a finite portion of an epoch; the succession of rock strata laid down within an age is called a stage.

apex predator: predator residing at the top of the food chain.

Arctoidea: mammalian superfamily containing dogs, weasels, and bears (**arctoideans**).

auditory bulla: bony structure containing parts of the middle and inner ear.

Barbourofelidae: Old World family of carnivorans closely related to the Nimravidae (**barbourofelids**).

basal: primitive.

Bergmann's rule: within a specific clade of animals, cooler climates favor larger body size whereas warmer climates favor smaller body size.

binomial: two-name combination given to a living or extinct species of animal or plant: the first is the genus name; the second is the specific epithet; and together they form the species name. Hence binomial nomenclature. Also called **binomen** (pl. **binomina**).

biotope: area of uniform environmental conditions providing a living place for a specific assemblage of plants and animals.

Bovidae: mammalian family that includes antelopes and cattle (**bovids**).

Calabrian: Mid- to late Pleistocene European land mammal age (1.81 to 0.78 Ma).

Caprini: bovid tribe that includes the sheep and goats (**caprins**).

carnassials: pair of teeth modified to shear flesh: upper fourth premolar (P4) and lower first molar (m1) in carnivorans, upper first molar (M1) and lower second molar (m2) in creodonts.

carnivoran: member of the order Carnivora.

carnivore: animal that eats flesh.

clade: group of organisms that evolved from a common ancestor.

cladistics: form of analysis that uses shared, unique characters to group organisms into clades.

cladogram: branching diagram showing the relationship between two or more species.

cline: continuous gradation from one population to another within a species.

competitive exclusion: elimination from a habitat of a species that is competing for the same resource as another species.

Creodonta: extinct order of carnivorous mammals (**creodonts**).

derived: evolved.

digitigrade: bearing the weight on the tips of the toes, like a dog or cat, so that the heel doesn't touch the ground.

endemic: restricted to a specific geographic region.

epoch: subdivision of a geologic period.

eutherian mammal: mammal whose female bears live young that are nourished within the body by a complex placenta (all mammals except marsupials and monotremes).

Felidae: mammalian family that includes living and extinct cats plus extinct sabertoothed cats (**felids**).

Felinae: mammalian subfamily that includes cats (**felines**).

Felini: tribe that includes most small cats (**felins**).

Gelasian: Early Pleistocene European land mammal age (2.58 to 1.81 Ma).

genome: complete set of genes or genetic material present in a cell or organism.

genus (plural genera): principal taxonomic category that ranks above species and below family; when written, the genus name starts with a capital letter and is presented in italics.

geochron: specific time based on geological divisions.

glacial: interval of time within an ice age that is characterized by lower temperatures and glacier advances.

guild: group of species that exploit the same resources or that exploit different resources in related ways.

Hemphillian: Late Miocene North American land mammal age (10.3 Ma to 4.9 Ma).

Herpestidae: family of carnivorans that includes the mongooses (**herpestids**).

holotype: specimen selected to exemplify a species when the species is first named.

Hyaenidae: family of carnivorans that includes the hyenas (**hyaenids**).

hypercarnivores: animal with a diet that is more than 70 percent meat.

incisors: teeth located anterior to the canines.

interglacial: interval of time within an ice age that is marked by warmer temperatures and the retreat of glaciers.

interstadial: short interval of warmer temperature within a glacial period.

ka: thousands of years.

keratin: fibrous protein that is the main constituent of fingernails, claws, hooves, hair, etc.

keystone species: plant or animal that plays a unique and crucial role in the way that the ecosystem functions.

land mammal age (LMA): interval of time defined by an assemblage of mammal species.

land mammal stage (LMS): succession of geological strata containing an assemblage that defines a land mammal age.

long branch attractions (LBA): form of systematic error whereby distantly related lineages are incorrectly inferred to be closely related.

Ma: millions of years.

Machairodontinae: cat subfamily that includes the sabertooths (**machairodontines**).

mammoth steppe: extensive Holarctic biome that, during glacial intervals, was characterized by high-productivity grasses, herbs, and willow and whose animal biomass was dominated by woolly mammoth, horse, and bison.

Mazegouan: Pliocene Asian land mammal age (3.6 to 2.59 Ma).

mesocarnivore: animal whose diet consists of 50–70 percent meat.

metapodials: bones that connect the wrist or ankle bones to the fingers or toes.

Miacidae: family of extinct carnivorous mammals thought to be the ancestors of carnivorans (**miacids**).

mitochondrial DNA (mtDNA): maternally inherited, extranuclear, double-stranded DNA found exclusively in mitochondria.

molars: teeth located behind the premolars.

molecular phylogenetics: branch of phylogeny that analyzes hereditary molecular differences, mainly in DNA sequences, to gain information about an organism's evolutionary relationships.

monophyletic: descended from a common evolutionary ancestor or ancestral group.

morphometrics: quantitative analysis of size and shape.

neofelids: true felids.

Nihewanian: Pleistocene Asian land mammal age (2.59 to 0.7 Ma).

Nimravidae: extinct family of mammalian carnivores, sometimes known as false saber-toothed cats (**nimravids**).

Ovibovini: bovid tribe that includes the muskox (**ovibovins**).

paleofelids: ancient cats or false cats.

Paleoryctidae: extinct family of relatively unspecialized eutherian mammals that arrived in North America during the Late Cretaceous and took part in the first placental evolutionary radiation (**paleoryctids**).

panther: leopard; sometimes used for a melanistic leopard or jaguar.

Panthera: genus of the Felidae that contains the extant roaring cats (lion, tiger, leopard, jaguar) and their immediate ancestors. For species of *Panthera* see appendix 4.

Pantherinae: subfamily containing the genus *Panthera* and closely related genera (**pantherines**).

Pantherini: tribe of big cats that contains *Panthera* and *Neofelis* (**pantherins**).

Paroodectes feisti: Middle Eocene **miacid** about the size of a housecat adapted for climbing and jumping in the tropical forests of the Proto-European archipelago.

period: subdivision of geologic time; the Earth's history is divided into twenty-two periods between 2500 Ma and the present.

phalanx: (plural **phalanges**): distal bones of the limb (fingers and toes).

phylogenetic: relating to the evolutionary development and diversification of a species or group of organisms or of a particular feature of an organism.

phylogeny: evolutionary history of a kind of organism.

plantigrade: stance with the sole of the foot lying flat on the ground.

premolars: teeth located between the canines and the molars.

preservation bias: factors that result in the preservation of anomalously large numbers of body parts or specimens.

Proailurus lemanensis: Early Miocene mammal, probably similar to the living fossa of Madagascar; one of the earliest mammals to show feline-like traits.

Pseudaelurus tournauensis (= *Pseudaelurus transitorius*): Early Miocene felid that could be a common ancestor of both the conical-toothed cats (subfamily Felinae) and the sabertoothed cats (subfamily Machairodontinae).

Rancholabrean: Late Pleistocene North American land mammal **age (190 ka to 11 ka)**.

refugium (plural **refugia**): area in which a population of organisms can survive during a period of otherwise unfavorable conditions.

riparian forest: forest adjacent to a body of water (river, lake, etc.).

Rupicaprini: bovid tribe that includes the mountain goat and chamois (**rupicaprins**).

Saigini: bovid tribe that includes the saiga antelope (**saigins**).

savanna: tropical or subtropical grassland with scattered trees.

sexual dimorphism: differences between males and females in size or other attributes.

shared characters: characters that two or more species have in common.

shear-bite: bite in which the sabertooth's flattened and keeled canines shear into flesh as the mouth is closed.

specific name: follows the genus name to form a species name or **binomen**; the specific name is written in italics and lower case (cf. **genus**).

stadial: short interval of cooler temperature within an interglacial period.

stage: succession of rock strata that accumulated during a geological age. By convention, a given age of rock and the corresponding age of time will have the same name and the same boundaries.

steppe: large area of flat, unforested grassland.

subgenus (plural **subgenera**): taxonomic category that ranks above species but below genus.

sympatric: occurring within the same geographic area.

systematics: the study of relationships.

taphonomy: study of what happens to an organism after death.

tundra: vast, flat, treeless arctic region in which the subsoil is permanently frozen.

veld: open uncultivated grassland in southern Africa.

Villanyian: Pliocene European land mammal age (3.6 to 2.58 Ma).

Viverridae: family of carnivorans that includes the civets (**viverrids**).

SOURCES

Adamson, J. 1960. *Born Free: a Lioness of Two Worlds.* New York: Pantheon.

Agustí, J., and M. Antón. 2002. *Mammoths, Sabertooths, and Hominids.* New York: Columbia University Press.

Akersten, W. A. 1985. "Canine Function in *Smilodon* (Mammalia; Felidae; Machairodontinae)." *Contributions in Science,* no. 356:1–22.

Antón, M. 2013. *Sabertooth.* Bloomington: Indiana University Press.

Antón, M., A. Galobart, and A. Turner. 2005. "Co-existence of Scimitar-Toothed Cats, Lions, and Hominins in the European Pleistocene: Implications of the Post-cranial Anatomy of *Homotherium latidens* (Owen) for Comparative Palaeoecology." *Quaternary Science Reviews* 24:1,287–301.

Argant, A., and J.-P. Brugal. 2017. "The Cave Lion *Panthera (Leo) spelaea* and Its Evolution: *Panthera spelaea intermedia* nov. Subspecies." *Acta Zoological Cracoviensia* 60(2): 61–104.

Attenborough, D. 1987. *The First Eden: The Mediterranean World and Man.* Boston,: Little, Brown and Company.

Baldwin, J. 1877. *The Large and Small Game of Bengal and the Northwestern Provinces of India.* London; Henry S. King & Co.

Bale, R. 2016. "The World Is Finally Getting Serious About Tiger Farms." *National Geographic Online,* https://news.nationalgeographic.com/2016/09/wildlife-watch-tiger-farms-cites-protections/.

Barnett, R., B. Shapiro, I. Barnes, et al. 2009. "Phylogeography of Lions (*Panthera leo* ssp.) Reveals Three Distinct Taxa and a Late Pleistocene Reduction in Genetic Diversity." *Molecular Ecology* 18:1,668–77.

Barnett, R., N. Yamaguchi, B. Shapiro, et al. 2014. "Revealing the Maternal Demographic History of *Panthera leo* Using Ancient DNA and a Spatially Explicit Genealogical Analysis." *BMC Evolutionary Biology* 14:70. doi:10.1186/1471-2148-14-70.

Baryshnikov, G. F. 2002. "Local Biochronology of Middle and Late Pleistocene Mammals from the Caucasus." *Russian Journal of Theriology* 1(1): 61–67.

Baryshnikov, G. F. 2011. "Pleistocene Felidae (Mammalia, Carnivora) from the Kudaro Paleolithic Cave Sites in the Caucasus." *Proceedings of the Zoological Institute RAS* 315(3): 197–226.

Baryshnikov, G. F. 2016. "Late Pleistocene Felidae Remains (Mammalia, Carnivora) from Geographical Society Cave in the Russian Far East." *Proceedings of the Zoological Institute RAS* 320(1): 84–120.

Baryshnikov, G. F., and G. Boeskorov. 2001. "The Pleistocene Cave Lion, *Panthera spelaea* (Carnivora, Felidae), from Yakutia, Russia." *Cranium* 18:7–24.

Benirschke, K.,S. Hiroaki, and T. Ito. 1972. "The Chromosomes of the Japanese Serow, *Capricornis crispus* (Temminek)." *Proceedings of the Japanese Academy* 48:608–12.

Biesty, S. 2003. *Rome in Spectacular Cross-Section.* New York: Scholastic Nonfiction.

Bocherens, H., D. G. Drucker, D. Bonjean, et al. 2011. "Isotopic Evidence for Dietary Ecology of Cave Lion (*Panthera spelaea*) in North-Western Europe: Prey Choice, Competition, and Implications for Extinction." *Quaternary International* 245:249–61.

Bogart, M., and K. Benirschke. 1975. "Chromosomes of a Male Takin (*Budorcas taxicolor taxicolor*)." *Chromosome Information Service* 16:18.

Borrero, L. A. 2001. *El problamiento de la Patagonia: Toldos, milodones y volcans.* Buenos Aires: Emecé.

Brain, C. K. 1969. "The Probable Role of Leopards as Predators of the Swartkrans Australopithecines." *South African Archaeological Bulletin* 24:170–71.

Bunch, T. D., C. Wu, Y.P. Zhang, and S. Wang. 2005. "Phylogenetic Analysis of Snow Sheep (*Ovis nivicola*) and Closely Related Taxa." *Journal of Heredity* 97(1): 21–30.

Burger, J., W. Rosendahl, O. Loreille, et al. 2004. "Molecular Phylogeny of the Extinct Cave Lion *Panthera leo spelaea*." *Molecular Phylogenetics and Evolution* 30:841–49.

Castelló, J. R. 2016. *Bovids of the World: Antelopes, Gazelles, Cattle, Goats, Sheep, and Relatives.* Princeton, NJ: Princeton University Press.

Chimento, N. R., and F. L. Agnolin. 2017. "The Fossil American lion (*Panthera atrox*) in South America": Palaeobiogeographical Implications." *Comptes Rendus Paleovol* 16:850–64.

Christiansen, P. 2007. "On the Distinctiveness of the Cape Lion (*Panthera leo melanochaita* Smith 1842) and a Possible New Specimen from the Zoological Museum, Copenhagen." *Mammalian Biology* 73:58–75.

Christiansen, P., and J. M. Harris. 2009. "Craniomandibular Morphology and Phylogenetic Affinities of *Panthera atrox*: Implications for the Evolution and Paleobiology of the Lion Lineage." *Journal of Vertebrate Paleontology* 29(3): 934–45.

Clottes, J. 2016. *What Is Paleolithic Art? Cave Paintings and the Dawn of Human Creativity.* Chicago: University of Chicago Press.

Cooper, D. L., A. J. Dugmore, B. M. Gittings, A. K. Scharf, A. Wilting, A. C. Kitchener. 2016. "Predicted Pleistocene–Holocene Range Shifts of the Tiger (*Panthera tigris*)." *Diversity and Distributions* 22:1,199–211.

Cueto, M., E. Camarós, P. Castaños, R. Ontañón, P. Arias. 2016. "Under the Skin of a Lion: Unique Evidence of Upper Paleolithic Exploitation and Use of

Cave Lion (*Panthera spelaea*) from the Lower Gallery of La Garma (Spain)." *PLOS One*. doi:10.1371/journal.pone.0163591.

Davis, B. W., G. Li., and W. J. Murphy. 2010. "Supermatrix and Species Tree Methods Resolve Phylogenetic Relationships Within the Big Cats, Panthera (Carnivora: Felidae)." *Molecular Phylogenetics and Evolution* 56(1): 64–76.

Day, D. 1981. *The Doomsday Book of Animals: A Natural History of Vanished Species*. New York: Viking Press.

Dayan, T., D. Simberloff, E. Tchernov, and Y. Yom-Tov. 1990. "Feline Canines: Community-Wide Character Displacement Among the Small Cats of Israel." *American Naturalist* 136 (1): 39–60.

Deng, T., X. Wang, M. Fortelius, et al. 2011. "Out of Tibet: Pliocene Woolly Rhino Suggests High-Plateau Origin of Ice-Age Herbivores." *Science* 333:1,285–88.

Diedrich, C. G. 2009. "Steppe Lion Remains Imported by Ice Age Spotted Hyenas Into the Late Pleistocene Perick Caves Hyena Den in Northern Germany." *Quaternary Research* 71:361–74.

Diedrich, C. G. 2010. "Spotted Hyena and Steppe Lion Predation Behaviours on Cave Bears of Europe—Late Quaternary Cave Bear Extinction as a Result Of Predator Stress." *Geophysical Research Abstracts* 12.

Diedrich, C. G., 2011a. "A Diseased *Panthera leo spelaea* (Goldfuss, 1810) Lioness from a Forest Elephant Graveyard in the Late Pleistocene (Eemian) Interglacial Lake at Neumark-Niord, Central, Germany." *Historical Biology* 23(2–3): 195–217.

Deidrich, C. G. 2011b. "The Largest European Cave Lion *Panthera spelaea* (Goldfuss 1810) Population from the Zoolithen Cave, Germany: Specialised Cave Bear Predators of Europe." *Historical Biology* 23:271–311.

Diedrich, C. G. 2011c. "Late Pleistocene *Panthera leo spelaea* (Goldfuss, 1810) Skeletons from the Czech Republic (central Europe): Their Pathological Cranial Features and Injuries Resulting from Intraspecific Fights, Conflicts with Hyenas, and Attacks on Cave Bears." *Bulletin of Geosciences* 86(4): 817–40.

Diedrich, C. G. 2013. "Late Pleistocene Leopards Across Europe—Northernmost European German Population, Highest Elevated Records in the Swiss Alps, Complete Skeletons in the Bosnia Herzegowina Dinarids and Comparison to the Ice Age Cave Art." *Quaternary Science Reviews* 76:167–93.

Diedrich, C. G. 2014. "Palaeopopulations of Late Pleistocene Top Predators in Europe: Ice Age Spotted Hyenas and Steppe Lions in Battle and Competition Over Prey." *Paleontology Journal*. 34 pages. doi:10.1155/2014/106203.

Divyabhanusinh, C. 2005. *The Story of Asia's Lions*. Mumbai: Marg.

Donlon, C. J., et al. 2006. "Pleistocene Rewilding: An Optimistic Agenda for Twenty-First Century Conservation." *American Naturalist* 168(5): 660–81.

Dubois, E. 1937. "On the Fossil Human Skulls Recently Discovered in Java and *Pithecanthropus erectus*." *Man* 37:1–29.

Ehlers, J., and P. L Gibbard. 2004. *Quaternary Glaciations—Extent and Chronology*. Vol. 2, part 1: *Europe*. Amsterdam: Elsevier Science.

Engel, M., J. J. Vaske, A. J. Bath, and S. Marchin. 2017. "Attitudes Toward Jaguar and Pumas and the Acceptability of Killing Big Cats in the Brazilian Atlantic Forest: An Application of the Potential for Conflict Index." *Ambio* 46(5): 604–12.

Ersmark, E., L. Orlando, E. Sandoval-Castellanos, et al. 2015. "Population Demography and Genetic Diversity in the Pleistocene Cave Lion." *Open Quaternary* 1(4): 1–14.

Fabrini, E. 1890. I "*Machairodus (Megantereon) del* Valdarno superior." *Bolletino del R. Comitato Geologico d'Italia* 21:121–44, 161–77.

Ficcarelli, G., and D. Torre. 1968. "Upper Villafranchian Panthers of Tuscany." *Palaeontographica Italica* 64:173–84.

Foronova, I. V. 2005. "Large Mammal Faunas from Southwestern Siberia of the Plio-Pleistocene Boundary and Lower/Middle Pleistocene Transition." *Quaternary International* 131:95–99.

Froese, D., M. Stiller, P. D. Heintzman, et al. 2017. "Fossil and Genomic Evidence Constrains the timing of Bison Arrival in North America." *Proceedings of the National Academy of Sciences* 114:3,457–62.

Gamble, C. et al. 2004. "Climate Change and Evolving Human Diversity in Europe During the Last Glacial." *Philosophical Transactions of the Royal Society of London Series B* 359:243–54.

Gavashelishvili, A., and V. Lukarevskiy. 2008. "Modelling the Habitat Requirements of Leopard *Panthera pardus* in West and Central Asia." *Journal of Applied Ecology* 45:579–88.

Goodyear, D. 2017. "Valley Cats: Are L.A.'s Mountain Lions Dangerous Predators or Celebrity Guests?" *New Yorker*, February 13 and 20.

Grayson, D. K., and D. J. Meltzer. 2003. "A Requiem for North American Overkill." *Journal of Anthropological Sciences* 30:585–93.

Grohé, C., B. Lee, and J. J. Flynn. 2018. "Recent Inner Ear Specialization for High-Speed Hunting in Cheetahs." *Scientific Reports* 8:2301. doi:10.1038/s41598-018-20198-3.

Gross, J. 2017. "The Snow Leopard: Ghost Cat of the Mountains." *Jaguar*, July 24. https://thejaguarandallies.com/2017/07/24/the-snow-leopard-ghost-cat-of-the-mountains/.

Guthrie, R. D. 1990. *Frozen Fauna of the Mammoth Steppe: The Story of Blue Babe.* Chicago: University of Chicago Press.

Hemmer, H., R. D. Kahlke, and T. Kelle. 2003. "*Panthera onca gombaszoegensis* (Kretzoi, 1938) from the Early Middle Pleistocene Mosbach Sands (Wiesbaden, Hessen, Germany)—a Contribution to the Knowledge of the Variability and History of the Dispersal of the Jaguar." *Neues Jahrbuch für Geologie und Paläontologie Abhandlung* 229(1): 31–60.

Hemmer, H., R-D. Kahlke, and A. K. Vakua. 2004. "The Old World Puma—*Puma pardoides* (OWEN, 1846) (Carnivora: Felidae)—in the Lower Villafranchian (Upper Pliocene) of Kvabebi (East Georgia, Transcaucasia) and Its Evolutionary and Biogeographical Significance." *Neues Jahrbuch für Geologie und Paläontologie Abhandlungen* 233(2): 197–231.

Hemmer, H., R-D. Kahlke, and A. K. Vakua. 2010. "*Panthera onca* ssp. nov. from the Early Pleistocene of Dmanisi (Republic of Georgia) and the Phylogeography of Jaguars (Mammalia, Carnivore, Felidae)." *Nues Jahrbuchj für Geologie und Palaontologie, Abhandlungen* 257(1): 115–27.

Heinrich, H. 1988. "Origin and Consequences of Cyclic Ice Rafting in the Northeast Atlantic Ocean During the Past 130,000 Years." *Quaternary Research* 29: 143–52.

Hoffecker, J. F., et al. 2016. "Beringia and the Global Dispersal of Modern Humans." *Evolutionary Anthropology* 25(2): 64–78.

Hooijer, D. A. 1947. "Pleistocene Remains of *Panthera tigris* (Linnaeus) Subspecies from Wahnsien, Szechwan, China, Compared with Fossil and Recent Tigers from Other Localities." *American Museum Novitates*, No. 1346:1–17.

Imbrie, J. A., A. Berger, E. A. Boyle, et al. 1993. "On the Structure and Origin Of Major Glaciation Cycles 2: The 100,000-Year Cycle." *Paleoceanography and Paleoclimatology* 8(6): 699–735. AGU 100. https://doi.org/10.1029/93PA02751.

Jacobson, A. P., P. Gerngross, J. R. Lemeris Jr. et al. 2016. "Leopard (*Panthera pardus*) Status, Distribution, and the Research Efforts Across Its Range." *PeerJ* 4:e1974. https://doi.org/10.7717/peerj.1974.

Janczewski, D. N., W. S. Modi, J. C. Stevens, and S. J. O'Brien. 1995. "Molecular Evolution of Mitochondrial 12S RNA and Cytochrome b Sequences in the Pantherine Lineage of Felidae." *Molecular Biology and Evolution* 12(4): 690–707.

Janecka, J. E., R. Jackson, Y. Zhang, et al. 2008. "Population Monitoring of Snow Leopards Using Noninvasive Collection of Scat Samples: A Pilot Study." *Animal Conservation* 11:401–11.

Jennison, G. 1937. *Animals for Show and Pleasure in Ancient Rome*. Manchester: Manchester University Press.

Kirillova, I. V., A. V. Tiunov, V. A. Levchenko, et al. 2015. "On the Discovery of a Cave Lion from the Malyi Anyui River (Chukotka, Russia)". *Quaternary Science Reviews* 117:135–51.

Kitchener, A. C. 1999. "Tiger Distribution, Phenotypic Variation, and Conservation Issues." In *Riding the Tiger: Tiger Conservation in Human-Dominated Landscapes*, ed. J. Seidensticke, S. Christie, and P. Jackson, 19–39. Cambridge: Cambridge University Press.

Kitchener, A. C., C. A. Driscoll, and N. Yamaguchi. 2016. "What Is a Snow Leopard? Taxonomy, Morphology, and Phylogeny." In *Snow Leopards—Biodiversity of the World: Conservation from Genes to Landscapes*, ed. D. McCarthy, T. Mellon, and P. Nyhus, 3–12. London: Academic Press.

Kitchener, A., and A. Dugmore. 2000. "Biogeographical Change in the Tiger, *Panthera tigris*." *Animal Conservation* 3(2): 113–24.

Kitchener, A., C. Breitenmoser-Würsten, E. Eizirik, et al. 2017. "A Revised Taxonomy of the Felidae: The Final Report of the Cat Classification Task Force of the IUCN/SSC Cat Specialist Group." *Cat News* 11, Special Issue, 1–80.

Kurtén, B. 1968. *Pleistocene Mammals of Europe*. Chicago: Aldine.

Kurtén, B. 1971. *The Age of Mammals*. New York: Columbia University Press.

Kurtén, B. 1972. *The Ice Age*. London: Rupert Hart-Davos.

Leakey, Richard E. 1981. *The Making of Mankind*. New York: E. P. Dutton.

Levy, S. 2011. *Once and Future Giants: What Ice Age Extinctions Tell Us About the Fate of the Earth's Largest Animals*. Oxford: Oxford University Press.

Li, Qiang, X. Wang, G. Xie, and A. Yin. 2013. "Oligocene-Miocene Mammalian Fossils from Hongyazi Basin and Its Bearing on Tectonics of Danghe Nanshan in Northern Tibetan Plateau." *PLOS One* 8:12e82816.

Li, Qiang, G. Xie, G. T. Takeucchi, et al. 2014. "Vertebrate Fossils on the Roof of the World: Biostratigraphy and Geochronology of High-Elevation Kunlun Pass Basin, Northern Tibetan Plateau, and Basin History as Related to the Kunlun Strike-Slip Fault." *Palaeogeography, Palaeoclimatology, Palaeoecology* 411:46–55.

Lioubine V. P. 1998. [The Acheulean epoch in the Caucasus]. *Archeologicheskie Studii* 47(1): 1–192 [in Russian].

Liu, L., J. T. Reonen, and M. Fortelius. 2009. "Significant Mid-latitude Aridity in the Middle Miocene of East Asia." *Palaeogeography, Palaeoclimatology, Palaeoecology* 279:201–6.

Luo, S.-J., J.-H. Kim, W. E. Johnson, et al. 2004. "Phylogeography and Genetic Ancestry of Tigers." *PLoS Biology* 2(12): 2,275–93.

MacDonald, D., and A. S. Loveridge. 2010. *The Biology and Conservation of Wild Felids.* Oxford: Oxford University Press.

Manthi, F. K., F. H. Brown, M. J. Plavcan, and L. Werdelin. 2017. "Gigantic Lion, *Panthera leo*, from the Pleistocene of Natodomeri, Eastern Africa." *Journal of Paleontology.* doi:10.1017/jpa.2017.68.

Martin, P. S. 1967. "Prehistoric Overkill." In *Pleistocene Extinctions: The Search for a Cause*, ed. P. S. Martin and H. E. Wright, 75–120. New Haven, CT: Yale University Press.

Masseti, M., and P. P. A. Mazza. 2013. "Western European Quaternary Lions: New Working Hypotheses." *Biological Journal of the Linnean Society* 109:66–77.

Matternes, J. H. 1985. "*Paranthropus robustus* Juvenile Cranium: Prey of Leopard 1.5 Million Years Ago." *National Geographic* (November).

Matthiessen, P. 2000. *Tigers in the Snow.* New York: North Point.

Mazák, J. H. 2010. "What Is *Panthera palaeosinensis?*" *Mammal Review* 40(1): 90–102.

Mazák, J. H., P. Christiansen, and A. C. Kitchener. 2011. "Oldest Known Pantherine Skull and Evolution of the Tiger." *PLoS ONE* 6(10): e25483. doi:10.1371/journal.pone.0025483.

McCrady, E. H., T. Kirby-Smith, and H. Templeton. 1951. "New Finds of Pleistocene Jaguar Skeletons from Tennessee Caves." *Proceedings of the United States National Museum* 101(3,287): 497–511.

McGowan, C. 1997. *The Raptor and the Lamb: Predators and Prey in the Living World.* New York: Henry Holt.

Miller, Daniel J. 1990. "Grasslands of the Tibetan Plateau." *Rangelands* 12(3):159–63.

Mills, J. A. 2015. *Blood of the Tiger: A Story of Conspiracy, Greed, and the Battle to Save a Magnificent Species.* Boston: Beacon.

Mishra, R. H., and M. Jeffries. 1991. *Royal Chitwan National Park.* Auckland: The Mountaineers/David Bateman.

Mosser, A. A., M. Kosmala, and C. Packer. 2015. "Landscape Heterogeneity and Behavioral Traits Drive the Evolution of Lion Group Territoriality." *Behavioral Ecology* 26(4): 1,051–59.

Nadler, C. F. 1971. "Chromosomes of the Dall Sheep, *Ovis dalli dalli* (Nelson)." *Journal of Mammalogy* 52(2): 461–63.

Nadler, C. F., R. S. Hoffmann, and A. Woolf. 1973. "G-band Patterns as Chromosomal Markers, and the Interpretation of Chromosomal Evolution in Wild Sheep (*Ovis*)." *Experientia* 29(1): 117–19.

Naish, D. 2008a. "Europe, Where the Sabretooths, Lions, and Leopards Are." *Tetrapod Zoology* http://scienceblogs.com/tetrapodzoology/2008/03/12/european-cats-part-i/.

Naish, D. 2008b. "Pumas of South Africa, Cheetahs of France, Jaguars of England." *Tetrapod Zoology.* http://scienceblogs.com/tetrapodzoology/2008/03/13/european-cats-part-ii/.

Nuwer, R. 2017. "Asia's Illegal Wildlife Trade Makes Tigers a Farm-to-Table Meal." *New York Times*, June 5. https://www.nytimes.com/2017/06/05/science /animal-farms-southeast-asia-endangered-animals.html.

Oken, L. 1816. *Lehrbuch der Naturgeschichte. Dritter Theil, Zoologie. Zweite Abtheilung, Fleischthier.* Jena: Schmidt.

O'Regan, H., and A. Turner. 2004. "Biostratigraphic and Palaeoecological Implications of New Fossil Felid Material from the Plio-Pleistocene Site of Tegelen, the Netherlands." *Palaeontology* 47(5): 1,181–93.

Pfeffer, P. 1968. *Asia: A Natural History.* New York: Random House.

Pobiner, B. 2008. "Paleoecological Information in Predator Tooth Marks." *Journal of Taphonomy* 6(3–4): 373–97.

Pushkina, D. 2006. "Dynamics of the Mammalian Fauna in Southern Siberia During the Late Paleolithic." *Vertebrata PalAsiatica* 3:262–73.

Pocock, R. I. 1917. "The Classification of Existing Felidae." *Annals and Magazine of Natural History*, Series 8, 20:329–50.

Qui, Z.-X. 2006. "Quaternary Environmental Changes and Evolution of Large Mammals in North China." *Chinese Academy of Sciences* 5: 1,000–3,118.

Rabanus-Wallace, T., M. J. Wooller, G. D. Zazula, et al. 2017. "Megafaunal Isotopes Reveal Role of Increased Moisture on Rangeland During Late Pleistocene Extinctions." *Nature Ecology & Evolution* 1(125). doi:10.1038/s41559-017-0125.

Rabinowitz, A. R. 1986. "Jaguar Predation on Domestic Livestock in Belize." *Wildlife Society Bulletin* 14:170–74

Rabinowitz, A. R. 1991a. "Behaviour and Movements of Sympatric Civet Species in Huai Kha Khaeng Wildlife Sanctuary, Thailand." *Journal of Zoology* 223:281–98.

Rabinowitz, A. R. 1991b. *Chasing the Dragon's Tail: The Struggle to Save Thailand's Wild Cats.* New York: Doubleday.

Rabinowitz, A. R. and B. G. Nottingham Jr. 1986. "Ecology and Behaviour of the Jaguar (*Panthera onca*) in Belize, Central America." *Journal of Zoology* 210:149–59.

Radinsky, L. 1969. "Outlines of Canid and Felid Evolution." *Annals New York Academy of Sciences* 167:277–88.

Roth, S., 1904. "Nuevos restos de mamíferos de la caverna Eberhardt en Última Esperanza." *Revista del Museo de La Plata* 11:37–52.

Sabol, M., J. Gullár, and J. Horvát. 2018. "Montane Record of the Late Pleistocene *Panthera spelaea* (Golfuss, 1810) from the Zapadne Tatry Mountains (Northern Slovakia)." *Journal of Vertebrate Paleontology* 38(3). doi:10.1080/02724634.2018 .1467921.

Sanderson, I. T. 1965. *Ivan T. Sanderson's Book of Great Jungles.* New York: Julian Messner.

Saragusty, J., A. Shavit-Meyrav, N. Yamaguchi, et al. 2014. "Comparative Skull Analysis Suggests Species-Specific Captivity-Related Malformation in Lions." *PLos One* 9(4): e94527. https//doi.org/10.1371/journal.pone.0094527.

Schaller, G. B. 1967. *The Deer and the Tiger: A Study of Wildlife in India.* Chicago: University of Chicago Press.

Schaller, G. B. 1972. *The Serengeti Lion: A Study of Predator-Prey Relations.* Chicago: University of Chicago Press.

Schaller, G. B., ed. 1977. *Mountain Monarchs: Wild Sheep and Goats of the Himalaya.* Chicago: Chicago University Press.

Schaller, G. B. 1982. *Stones of Silence: Journeys in the Himalaya.* New York: Bantam.

Schaller, G. B. 1998. *Wildlife of the Tibetan Steppe.* Chicago: Chicago University Press.

Schaller, G. B., 2012. *Tibet Wild: A Naturalist's Journeys on the Roof of the World.* Washington, DC: Island.

Seidensticker, J. 2010. "Saving Wild Tigers: A Case Study in Biodiversity Loss and Challenges to Be Met in Recovery Beyond 2010." *Integrative Zoology* 5(4): 285–99.

Seymour, K. L. 1989. "*Panthera onca.*" *Mammalian Species,* no. 340 (October): 1–9.

Shackleton, N., M. Hall, and D. Pate. 1995. "Pliocene Stable Oxygen and Carbon Isotope Record of Benthic Foraminifera from ODP Site 138-846 in the Eastern Equatorial Pacific." *Pangea.* https://doi.pangaea.de/10.1594/PANGAEA.696450.

Shapiro, B. 2015. *How to Clone a Mammoth: The Science of De-Extinction.* Princeton, NJ: Princeton University Press.

Shi, Qi. 2013. "New Species of *Tsaidamotherium* (Bovidae, Artiodactyla) from China Sheds New Light on the Skull Morphology and Systematics of the Genus." *Science China Earth Sciences* 57(2):258–66.

Shu-Jin, L., J.-H. Kim, W. E Johnson, et al. 2004. "Phylogeography and Genetic Ancestry of Tigers (*Panthera tigris*)." *PLOS Biology* 2(12): e442. doi:10.1371/journal.pbio.0020442.

Sorkhabi, R. B., and E. Stump. 1993. "Rise of the Himalayas: A Geochronologic Approach." *GSA Today* 3:485–91.

Sotnikova, M., and P. Nikolsky. 2006. "Systematic Position of the Cave Lion *Panthera spelaea* (Goldfuss) Based on Cranial and Dental Characters." *Quaternary International* 142–43:218–28.

Sotnikova, M. V., and I. V. Foronova. 2014. "First Asian Record of *Panthera* (*Leo*) *fossilis* in the Early Pleistocene of Western Siberia, Russia." *Integrative Zoology* 9:517–30.

Soulé, M. E., and R. K. Noss. 1998. "Rewilding and Biodiversity as Complementary Tools for Continental Conservation." *Wild Earth* (Fall): 18–28.

Spassov, N. 1998. "A New Late Villafranchian Locality of Vertebrate Fauna—Slivnitsa (Bulgaria) and the Carnivore Dispersal Events in Europe on the Pliocene/Pleistocene Boundary." *Historia Naturalis Bulgarica* 9:101–13.

Stuart, A. J., and A. M. Lister. 2007. "Patterns of Late Quaternary Megafaunal Extinctions in Europe and Northern Asia." *Courier Forschungsinstitut Senckenberg* 259:1,287–2,971.

Stuart, A. J. and A.M. Lister. 2011. "Extinction Chronology of the Cave Lion *Panthera spelaea.*" *Quaternary Science Reviews* 30:2,329–40.

Switek, B. 2011. "American Lion, or Giant Jaguar?—in Search of *Panthera atrox.*" *Wired,* October 24. https://www.wired.com/2011/10/american-lion-or-giant-jaguar-in-search-of-panthera-atrox/.

Teague, R. L., 2009. "The Ecological Context of the Early Pleistocene Hominin Dispersal to Asia." Ph.D. diss, George Washington University.

Toomey, D. 2015. "How Tiger Farming in China Threatens the World's Tigers." *Yale Environment 360,* January 20. http://e360.yale.edu/features/how_tiger_farming_in_china_threatens_worlds_wild_tigers.

Tseng, Z. J., X. Wang, G. J. Slater, et al. 2014. "Himalayan Fossils of the Oldest Known Pantherine Establish Ancient Origin of Big Cats." *Proceedings of the Royal Society*, Series B, 281: 2,013–686. doi:10.1098/rspb.2013.2686.

Turner, A., and M. Antón. 1997. *The Big Cats and Their Fossil Relatives.* New York: Columbia University Press.

Turner, A., and M. Antón. 1998. "Climate and Evolution: Implications of Some Extinction Patterns in African and European Machairodontine Cats of the Plio-Pleistocene." *Estudios Geológicos* 54:209–30.

Turner, A., and M. Antón. 2004. *Evolving Eden: An Illustrated Guide to the Evolution of the Large African Mammal Fauna.* New York: Columbia University Press.

Ukge, 2016. "Fossil Lions of Europe." *Deposits*, February 18. https://depositsmag .com/2016/02/18/fossil-lions-of-europe/.

Valdez, F. P., T. Haag, F. C. C. Azevedo, et al. 2015. "Population Genetics of Jaguars (*Panthera onca*) in the Brazilian Pantanal: Molecular Evidence for Demographic Connectivity on a Regional Scale." *Journal of Heredity* 106(S1): 503–11.

Van Orsdol, K., J. P. Hanby, and J. D. Bygott. 1985. "Ecological Correlates of Lion Social Organization (*Panthera leo*)." *Journal of Zoology (A)* 206: 97–112.

Van Valkenburgh, B. 1989. "Carnivore Dental Adaptations and Diet: A Study of Trophic Diversity Within Guilds." In *Carnivore Behavior, Ecology, and Evolution,* ed. J. L. Gittelman, 410–36. Boston: Springer.

Van Valkenburgh, B. 1993. "Tough Times At La Brea: Tooth Breakage in Large Carnivores of the Late Pleistocene." *Science* 261(5,120): 456–59.

Van Valkenburgh, B. 1996. "Feeding Behavior in Free-Ranging, Large African Carnivores." *Journal of Mammalogy* 77(1): 240–54.

Van Valkenburgh, B., et al. 2016. "The Impact of Large Terrestrial Carnivores on Pleistocene Ecosystems." *Proceedings of the National Academy of Sciences* 113(4): 862–67.

Velichko, A. A., A. A. Andreev, and V. A. Klimanov. 1997. "Climate and Vegetation Dynamics in the Tundra and Forest Zone During the Late Glacial and Holocene." *Quaternary International* 41–42:71–96.

Vislobokova, I., and G. Daxner-Hock. 2002. "Oligocene–Early Miocene Ruminants from the Valley of Lakes (Central Mongolia)." *Annals of the Naturalhistoriche Museum of Wien* 103:213–35.

Volmer, R., C. Hertler, and A. van der Geer. 2015. "Niche Overlap and Competition Potential Among Tigers (*Panthera tigris*), Sabertoothed Cats (*Homotherium ultimum, Hemimachairodus zwierzyckii*), and Merriam's Dog (*Megacyon merriami*) in the Pleistocene of Java." *Palaeogeography, Palaeoclimatology, Palaeoecology* 441(4): 901–11. doi.10.1016/j.palaeo.2015.10.039:0031-0182.

Vrba, E. S., and G. B. Schaller, eds. 2000. *Antelopes, Deer and Relatives: Fossil Record, Behavioral Ecology, Systematics, and Conservation.* New Haven, CT: Yale University Press.

Wang, C., X. Zhao, Z. Liu, et al. 2008. "Constraints on the Early Uplift History of the Tibetan Plateau." *Proceedings of the National Academy of Sciences* 105(13): 4,987–92.

Wang, E. 2015. "Flexural Bending of Southern Tibet in a Retro Foreland Setting." *Nature Scientific Reports* 5:12076. doi:10.1038/srep12076.

Wang, Xiaoming., L. J. Flynn, and M. Fortelius, eds. 2013. *Fossil Mammals of Asia: Neogene Biostratigraphy and Chronology.* New York: Columbia University Press.

Wang, Xiaoming., Q. Li, and G. Takeuchi. 2016. "Out of Tibet: An Early Sheep from the Pliocene of Tibet, *Protovis himalayensis,* gen. et sp. nov. (Bovidae, Caprini), and Origin of Ice Age Mountain Sheep." *Journal of Vertebrate Paleontology* 36(5): e1169190.

Wang, Xiaoming., Z. Qiu, Q. Li, et al. 2007. "Vertebrate Paleontology, Biostratigraphy, Geochronology, and Paleoenvironment of Qaidam Basin in Northern Tibetan Plateau." *Palaeogeography, Palaeoclimatology, Palaeoecology* 254:363–85.

Wang, Xiaoming., Y. Wang, Q. Li, et al. 2015. "Cenozoic Vertebrate Evolution and Paleoenvironment in Tibetan Plateau: Progress and Prospects." *Gondwana Research* 27:1,335–54.

Wang, Xiao Xian, J.J. Zhang, J. Liu, S.Y. Yan, and J. M. Wang. 2013. "Middle-Miocene Transformation of Tectonic Regime in the Himalayan Orogen." *Chinese Science Bulletin* 58(1): 108–77.

Wang, Y., T. Deng, and D. Biasatti. 2006. "Ancient Diets Indicate Significant Uplift of Southern Tibet After ca. 7 Ma." *Geology* 34(4): 309–12.

Wen, J., J.-Q. Zhang, Z.-L. Nie, Y. Zhong, and H. Sun. 2014. "Evolutionary Diversifications of plants on the Qinghai-Tibetan Plateau." *Frontiers in Genetics* 5:4.

Werdelin, L., and M. E. Lewis. 2005. "Plio-Pleistocene Carnivora of Eastern Africa: Species Richness and Turnover Patterns." *Zoological Journal of the Linnean Society* 144:121–44.

Werdelin, L., and M. E Lewis. 2013. "Temporal Change in Functional Richness and Evenness in the Eastern African Plio-Pleistocene Carnivoran Guild." *PLoS One* 8(3): e57944.

Werdelin, L., N. Yamaguchi, W. E. Johnson, and S. J. O'Brien. 2010. "Phylogeny and Evolution of Cats." In *The Biology and Conservation of Wild Felids*, ed. D. MacDonald, and A. S. Loveridge, 59–82. Oxford: Oxford University Press.

Wheeler, T. D., and G. T. Jefferson. 2009. "*Panthera atrox*: Body Proportions, Size, Sexual Dimorphism, and Behavior of the Cursorial Lion of the North American Plains." In *Papers on Geology, Vertebrate Paleontology, and Biostratigraphy in Honor of Michael O. Woodburne*, ed. L. B. Albright III, 423–44. Museum of Northern Arizona Bulletin 65, Flagstaff: Museum of Northern Arizona.

White, P. A., and C. G. Deidrich. 2012. "Taphonomy Story of a Modern African Elephant *Loxodonta africana* Carcass on a Lakeshore in Zambia (Africa)." *Quaternary International* 247:287–96.

Wurster, D. H., and K. Benirschke. 1968. "Chromosome Studies in the Superfamily Bovoidea." *Chromosoma* 25(2): 152–71.

Yang, S., H. Dong, and F. Lei. 2009. "Phylogeography of Regional Fauna on the Tibetan Plateau: A Review." *Progress in Natural Science* 19:789–99.

Environmental Investigation Agency. 2019. "Panthera, EIA, Wildlife Conservation Trust Urge WHO to Condemn Traditional Chinese Medicine Utilizing Wild Animal Parts. *YubaNet.com.* May 20. https://yubanet.com/world/panthera-eia-wildlife-conservation-trust-urge-who-to-condemn-traditional-chinese-medicine-utilizing-wild-animal-parts/.

Zazula, Grant D., D. G. Froese, C. E. Schweger, et al. 2003. "Ice-Age Steppe Vegetation in East Beringia." *Nature* 423:603. doi:10.1038/423603a.

Zhang, Q., F. Wang, H. Ji, and W. Huang. 1981. "Pliocene Sediments of the Zanda Basin, Tibet." *Journal of Stratigraphy* 5(3): 216–20.

SUGGESTED READING

Adamson, J. 1960. *Born Free: A Lioness of Two Worlds.* New York: Pantheon Books,.

Agustí, J., and M. Antón. 2002. *Mammoths, Sabertooths and Hominids.* New York: Columbia University Press.

Antón, M. 2013. *Sabertooth.* Bloomington: Indiana University Press.

Attenborough, D. 1987. *The First Eden: The Mediterranean World and Man.* Boston: Little, Brown.

Castelló, J. R. 2016. *Bovids of the World: Antelopes, Gazelles, Cattle, Goats, Sheep, and Relatives.* Princeton, NJ: Princeton University Press.

CITES: Convention on International Trade in Endangered Species of Wild Fauna and Flora. https://www.cites.org/eng.

Day, D. 1981. *The Doomsday Book of Animals: A Natural History of Vanished Species.* New York: Viking.

Divyabhanusinh, C. 1995. *The End of the Trail.* Delhi: Banyen Books.

Divyabhanusinh, C. 2005. *The Story of Asia's Lions.* Mumbai: Marg Publications.

Donlon, C. J., et al. 2006. "Pleistocene Rewilding: A Optimistic Agenda for Twenty-First Century Conservation." *American Naturalist* 168 (5): 660–81.

Jennison, G. 1937. *Animals for Show and Pleasure in Ancient Rome.* Manchester: Manchester University Press.

Kurtén, B. 1968. *Pleistocene Mammals of Europe.* Chicago: Aldine.

Kurtén, B. 1971. *The Age of Mammals.* New York: Columbia University Press.

Kurtén, B. 1972. *The Ice Age.* London: Rupert-Hart Davos.

Leakey, R. E. 1981. *The Making of Mankind.* New York: E. P. Dutton.

Levy, S. 2011. *Once and Future Giants: What Ice Age Extinctions Tell Us About the Fate of the Earth's Largest Animals.* Oxford: Oxford University Press.

MacDonald, D., and A. S. Loveridge. 2010. *The Biology and Conservation of Wild Felids.* Oxford: Oxford University Press.

Matthiessen, P. 2000. *Tigers in the Snow.* New York: North Point Press.

McGowan, C. 1997. *The Raptor and the Lamb: Predators and Prey in the Living World.* New York: Henry Holt.

Mills, J. A. 2015. *The Blood of the Tiger: A Story of Conspiracy, Greed, and the Battle to Save a Magnificent Species.* Boston: Beacon.

Pfeffer, P. 1968. *Asia: A Natural History.* New York: Random House.

Rabinowitz, A. R., 1986. *Jaguar: One Man's Struggle to Establish the World's First Jaguar Preserve.* New York: Arbor House.

Rabinowitz, A. R. 1991. *Chasing the Dragon's Tail: The Struggle to Save Thailand's Wild Cats.* New York: Doubleday.

Schaller, G. B. 1967. *The Deer and the Tiger.* Chicago: University of Chicago Press.

Schaller, G. B. 1972. *The Serengeti Lion: A Study of Predator-Prey Relations.* Chicago: University of Chicago Press.

Schaller, G. B. 1977. *Mountain Monarchs: Wild Sheep and Goats of the Himalaya.* Chicago: University of Chicago Press.

Schaller, G. B. 1982. *Stones of Silence: Journeys in the Himalaya.* New York: Bantam.

Schaller, G. B. 1998. *Wildlife of the Tibetan Steppe.* Chicago: University of Chicago Press.

Schaller, G. B. 2012. *Tibet Wild: A Naturalist's Journeys on the Roof of the World.* Washington, DC: Island Press.

Shapiro, B. 2015. *How to Clone a Mammoth: The Science of De-Extinction.* Princeton, NJ: Princeton University Press.

Turner, A., and M. Antón. 1997. *The Big Cats and Their Fossil Relatives.* New York: Columbia University Press.

Turner, A., and M. Antón. 2004. *Evolving Eden: An Illustrated Guide to the Evolution of the Large African Mammal Fauna.* New York: Columbia University Press.

Vrba, E. S., and G. B. Schaller, eds. 2000. *Antelopes, Deer, and Relatives: Fossil Record, Behavioral Ecology, Systematics, and Conservation.* New Haven, CT: Yale University Press.

INDEX

Page numbers in *italics* refer to figures.

Printed in the USA
CPSIA information can be obtained
at www.ICGtesting.com
JSHW051448221024
72173JS00007B/1610

9 780231 184502